Marx Joyce
Abbott Hardy Austen
Defoe Melville Cooper Hugo
Machiavelli Chesterton Eliot
Montaigne Emerson Grimm
Stoker Christie Haggard
Carroll Molière
Wilde Maupassant Byron
Garnett Einstein Engels Schiller
Goethe Fitzgerald Hawthorne Kafka
Cotton Dostoyevsky Smith
Baum Kipling Doyle Hall
Leslie Henry Willis
Dumas Flaubert Nietzsche
Stockton Turgenev Balzac
Burroughs Vatsyayana Crane
Curtis Tocqueville Verne
Homer Whitman Gogol Vinci
Darwin Widger Tolstoy Busch
Potter Thoreau
Freud Zola Twain Scott
Kant Jowett Lawrence Plato Harte
Stevenson Dickens
Andersen Hesse
London Descartes Cervantes Burton
Poe Aristotle Voltaire
Hale James Hastings Cooke
Bunner Shakespeare Irving
Richter Chambers
Doré da
Dante Shaw Benedict
Swift Chekhov Wodehouse Alcott
Pushkin
Newton

tredition®

tredition was established in 2006 by Sandra Latusseck and Soenke Schulz. Based in Hamburg, Germany, tredition offers publishing solutions to authors and publishing houses, combined with worldwide distribution of printed and digital book content. tredition is uniquely positioned to enable authors and publishing houses to create books on their own terms and without conventional manufacturing risks.

For more information please visit: www.tredition.com

TREDITION CLASSICS

This book is part of the TREDITION CLASSICS series. The creators of this series are united by passion for literature and driven by the intention of making all public domain books available in printed format again - worldwide. Most TREDITION CLASSICS titles have been out of print and off the bookstore shelves for decades. At tredition we believe that a great book never goes out of style and that its value is eternal. Several mostly non-profit literature projects provide content to tredition. To support their good work, tredition donates a portion of the proceeds from each sold copy. As a reader of a TREDITION CLASSICS book, you support our mission to save many of the amazing works of world literature from oblivion. See all available books at www.tredition.com.

 Project Gutenberg

The content for this book has been graciously provided by Project Gutenberg. Project Gutenberg is a non-profit organization founded by Michael Hart in 1971 at the University of Illinois. The mission of Project Gutenberg is simple: To encourage the creation and distribution of eBooks. Project Gutenberg is the first and largest collection of public domain eBooks.

Domesticated Animals Their Relation to Man and to his Advancement in Civilization

Nathaniel Southgate Shaler

Imprint

This book is part of TREDITION CLASSICS

Author: Nathaniel Southgate Shaler
Cover design: Buchgut, Berlin – Germany

Publisher: tredition GmbH, Hamburg - Germany
ISBN: 978-3-8472-2160-9

www.tredition.com
www.tredition.de

Copyright:
The content of this book is sourced from the public domain.

The intention of the TREDITION CLASSICS series is to make world literature in the public domain available in printed format. Literary enthusiasts and organizations, such as Project Gutenberg, worldwide have scanned and digitally edited the original texts. tredition has subsequently formatted and redesigned the content into a modern reading layout. Therefore, we cannot guarantee the exact reproduction of the original format of a particular historic edition. Please also note that no modifications have been made to the spelling, therefore it may differ from the orthography used today.

African Elephant

DOMESTICATED ANIMALS
THEIR RELATION TO MAN AND TO HIS
ADVANCEMENT IN CIVILIZATION

BY

NATHANIEL SOUTHGATE SHALER
DEAN OF THE LAWRENCE SCIENTIFIC SCHOOL OF
HARVARD UNIVERSITY

ILLUSTRATED

NEW YORK
CHARLES SCRIBNER'S SONS
1908
Copyright, 1895, by
CHARLES SCRIBNER'S SONS

DOMESTICATED ANIMALS

INTRODUCTION

One of the effects of the modern advance in natural science has been greatly to increase the attention which is devoted to the influences that the conditions of diverse peoples have had upon their development. Man is no longer looked upon, as he was of old, as a being which had been imposed upon the earth in a sudden and arbitrary manner, set to rule the world into which he had been sent as a master. We now see him as one of the myriad species which has won its way by powers of mind out of darkness and the great struggle to the place of command. The way in which this creature, weak in body and exceedingly dependent on his surroundings, has in the modern geologic epoch come forth from the mass of the lower animals, is by far the most impressive and as yet the most unexplained phenomenon which the geologist has to consider. It is not likely that the marvellous advancement can be accounted for by any single cause; it is probably due, as are most of the great evolutions, to the concurrence of many influences; but among these which make for advance, we clearly have to reckon the animals and plants which man has learned to associate with his work of the household and the fields.

Although certain species of insects, particularly the ants, have the well-developed habit of subjugating certain creat [2] ures of their own family, man is the only vertebrate that has ever adopted the plan of domesticating a variety of animals and plants. The beginnings of this custom were made in a very remote time, and for long ages the profit which was thereby gained appears to have been but slight. Gradually, however, races, owing to their masterful quality and to the opportunities which were offered by the wild life about their dwelling places, obtained flocks and herds. In the group of continents commonly termed the old world, where there were several ancient primitive peoples of innate ability, and where there were many species of larger mammals which were well fitted for domestication, the advance in social development went on rapidly. In the new world, though the primitive races contained tribes of much ability, there was practically no chance for the people to add to their strength by the subjugation of beasts of burden, or to their food resources by the adoption of various animals which could be used for the needs of food or raiment. The advance of men when they have obtained valuable domesticated animals, and their failure to win a high station where the surrounding nature denied such opportunities, go far to prove the bearing of this accomplishment in the development of peoples.

A little consideration makes it evident to us that the advance of mankind above the original savage state is in several ways favored by the possession of domesticated animals. In the first place, each creature which is adopted into the household or the fields usually brings as its tribute a substantial contribution to the resources which tend to make the society commercially successful. When we consider the enlargements of resources and the diversification of indus [3] tries which rest upon the adoption of any one of these animals—as, for instance, the horse—we see in a way what the possession of domesticated animals and plants really means, and are in a position to conceive, though at best but dimly, what the scores of these captive species have done for us. We recognize the fact that while, under almost any conditions, a certain manner of advance above the most primitive savagery is possible to a naturally able people, this on-going cannot lead any distance unless the folk have other help than their own weak bodies can give them. It is hardly too much to say that civilization has intimately depended on the subjugation of a great range of useful species.

It would be interesting to trace, if we could, what share the several domesticated animals have had in the development of the human races; but this task is not to be done. We can, however, discern that the Arab without the camel and the horse would not have found the place in history which he has filled, and that our own race could not have attained its place save for the aid which the horned cattle, sheep, and a host of other helpers which we have pressed into service, have afforded. These economic gains have to be judged in mass, they cannot be reckoned in detail. When we have made the best account of them we can, there remains another class of influences, the value of which, though evidently great, is yet harder to reckon; these arise from the education which has been attained through the care of these adopted creatures. Among savages the great need is a training in forethoughtfulness; all primitive peoples are like children, they live in the interests of the day; the cares of the seasons to come, or even of the morrow, are not for them. The possession of domesticated animals certainly did [4] much to break up this old brutal way of life; it led to a higher sense of responsibility to the care of the household; it brought about systematic agriculture; it developed the art of war; it laid the foundations of wealth and commerce, and so set men well upon their upward way. Moreover, the use of domesticated animals of the better sort enabled the more vigorous and care-taking races to gain the strength which led to their advancement in power to a point where they were able to displace the lower and feebler tribes. In other words, the system of domestication has provided a method by which those peoples who were fitted to develop the qualities which make for civilization could advance; it has provided the opportunity for selection.

Of all the influences which have been exercised on man by the care of his flocks, herds, and droves, perhaps the most important is that which has arisen from the broader development of his sympathies. The savage may be defined as a man who cares only for his family and his tribe; the civilized man as one whose kindly interest extends to mankind and beyond to all sentient beings. In the development of this altruistic motive the care of the dependent species has evidently been most effective. We note that the peoples who have attained the first upward step in the association with domesticated animals are in their quality, so far as tested by literature and history, much above the mere savage. With the care of the flocks we find associated poetry, the first notes of higher religious motives, and a largeness of the sympathetic life which is favored by the nature of the occupation. Where the nomadic habits of the original shepherds pass into the more sedentary state of the soil tiller, the element of personal care and the affection and [5] the consequent education of the sympathy were increased. Men had now to care for half a dozen or more kinds of animals; they had to learn their ways, in a manner to put themselves in their places and conceive their needs. Thus the life of a farmer is a continual lesson in the art of sympathy; with the result, certainly in part due to this cause, that there is no class of people from whom the brutal instincts of the ancient savage life which we all inherit have been so completely eradicated.

It is perhaps too much to attribute the advance of the agricultural classes of our civilized peoples, in all that serves to remove them from the brutality of their savage ancestors, altogether to the nature of their work—to the very large element of kindly care for which it calls, and which is the price of success in the occupation. Yet when we note the immediate way in which the people bred in cities, under circumstances of excitement are wont to behave like savages of the lower kind, showing in their conduct a lack of all sympathetic education, and contrast their behavior with that of their kinsmen from the fields—we see essential differences in character which cannot well be explained save by the diverse natures of the training which the men have received. Thus in the French Revolution, the baser, more inhuman deeds were not committed by the peasants, who had been the principal sufferers under the régime which was overthrown, but by the people of the great towns who had been less oppressed by the iniquities of the old system of government.

If it be true—as my personal experiences and observations lead me firmly to believe is the case—that man's contact with the domesticated animals has been and is ever to be one of [6] the most effective means whereby his sympathetic, his civilized motives may be broadened and affirmed, there is clearly reason for giving to this side of life a larger share of attention than it has received. So far the presence of these lower creatures in our society has generally been accepted as a matter of course. Sentimentalists, after the fashion of Laurence Sterne, have dwelt upon the imaginary woes of the creatures. Associations of well-meaning people have endeavored to diminish the cruelty which people of the towns, rarely those bred on the soil, often inflict upon them. It seems, however, desirable that we should place this consideration upon a plane more fitting the knowledge of our time. It should be made plain, not only that the success of our civilization depends now as in the past on the coöperation which mankind has had from the domesticated animals, but also that the development of this relation is one of the most interesting features in all history. On through the ages of the geologic past comes this great procession of life, in the endless succession of species whose numbers in the aggregate are to be reckoned by the scores, if not by the hundreds of millions. Until this modern age, the throng goes forward blindly, groping its way towards the higher planes of life. At length certain of the more advanced forms attain to a measure of intellectual elevation. Still, for all this advance, the life is not organized so as to attain any large ends; no society arises from it.

Suddenly, in the last geological epoch, man, the descendant of a group which like all others had led the narrow life of the preparatory ages, appears upon the scene. At first, and in his lower human estate, his position was not noticeably higher than that of his kindred, but there was in him [7] the seed of a great unlikeness, of very new things, in that his desires had an element of the unlimited which was to grow apace, and in time to make him greedy of ongoing. As this innovating creature sought for agents of power in the wilderness about him, he blindly laid hands upon such of the fellow tenants of the wilds as might serve his immediate needs. This species, both animals and plants, endowed with the capacity for variation, the plasticity which is in general a characteristic of all organic forms, were early led by their new master, as of old they had been guided by the old organic laws. They changed according to his choice, abandoning their ancient ways for the novel paths of civilization. With this association of the higher forms of the earth under the leadership of man, there began an entirely new and unprecedented condition of the world's affairs. In place of the ancient law of nature there came the control of our species which had been, in a way, chosen to be the overlord of life.

At first, the number of species of animals and plants which man brought under his control was very limited; it was indeed confined to those which might readily be subjugated to meet immediate needs. Gradually, however, the list has been extended until it included thousands of forms, which, while they meet no need such as the savage recognizes, are gratifying to the taste or the ambitions of civilized peoples. These æsthetic devices, or those of necessity, are advancing so rapidly that each generation sees hundreds of new animal and plant species added to our living collections, so that our plant and animal gardens now contain a large share of the more attractive forms which are to be found in the various geographical realms. Our tilled fields yield perhaps a [8] hundred times as many varieties of plants as they did in the earliest historic agriculture. The advance in the process of domestication is not so rapid as regards the animal kingdom as it is with the realm of plants, and this mainly for the reason that animals have a will of their own which has to be bent or broken to that of man. Still it goes on apace. We of to-day have at our command many times the number of sentient species contributive to our pleasure or profit that had been made captive at the beginning of our era. Naturally, in the early days of domestication, men brought under their control the greater number of the animals which gave promise of utility. As no new species of any economic importance have been created within the last geologic period, the field for the extension of economic domestication has of late been very limited. But the realm of sympathetic appreciation, unlike the economic, knows no definite bounds, and promises in time to bring all the more important organic forms under the care of the sympathetic and masterful being who has been chosen as the ruler of terrestrial life.

We thus see that the matter of domesticated animals is but a part of the larger problem which includes all that relates to man's destined mastery of the earth—a mastery which he is rapidly winning. It means that, in time, a large part of the life of this sphere is to be committed to his care, to survive or perish as he wills, to change at his bidding, to give, as other subjugated kinds have done, whatever of profit or pleasure they may contribute to his endless advancement. From this point of view our domesticated creatures should be presented to our people, with the purpose in mind of bringing them to see that the process of domestication has a far-reaching aspect, a dignity, we may fairly say a grandeur, that [9] few human actions possess. If we can impress this view, it will be certain to awaken men to a larger sense of their responsibility for, and their duty by, the creatures which we have taken from their olden natural state into the social order. It will, at the same time, enlarge our conceptions of our own place in the order of this world.

In the following pages little effort has been made to present those facts concerning domesticated animals which would commonly be reckoned as scientific. The several essays which, in larger part, were separately printed in Scribner's Magazine, are intended for those persons who, while they may not care to approach the matter in the manner of the professional inquirer, are glad to have the results which naturalists have attained, so far as they may serve to extend knowledge of things which lie in the field of familiar experiences. To the text as it at first appeared, numerous additions have been made, and the concluding chapters, on the Rights of Animals, and on the Problem of Domestication, are new. In them an effort is made to direct attention to the importance of the problem of man's relation to the lower life which is about him, and which in the future far more than in the past is to be helped or hindered by his rule. Our life is made up of large problems; but there seem few that are greater than this, which concerns our duty by the creatures that share with us the blessings of existence, and over which we have come to rule.

Sheep-Dogs Guarding a Flock at Night

THE DOG

Ancestry of the Domesticated Dogs. — Early Uses of the Animal: Variations induced by Civilization. — Shepherd-dogs: their Peculiarities; other Breeds. — Possible Intellectual Advances. — Evils of Specialized Breeding. — Likeness of Emotions of Dogs to those of Man: Comparison with other Domesticated Animals. — Modes of Expression of Emotions in Dogs. — Future Development of this Species. — Comparison of Dogs and Cats as regards Intelligence and Position in Relation to Man.

It is an interesting fact that the first creature which man won to domesticity was made captive and friend for the sake of companionship rather than for any grosser profit. The dog was, the world over, the first living possession of man beyond the limits of his own kindred. He has been so long separated from the primitive species whence he sprang that we cannot trace with any certainty his kinship with the creatures of the wilderness. Like his master he has become so artificialized that it is hard to conjecture what his original state may have been.

Naturalists are much divided in opinion in all that relates to the origin of our ancient and common domesticated animals; and this for the reason that the longer a creature has been subjected to the change-bringing conditions of our fields and households, the further it has departed from the parent stock. This difficulty is naturally the greatest in the case of the dogs, for the reason that they have been longer and more completely under the control of man than any other of the lower animals. Some students of the problem have inclined to the opinion that the dog is a descendant [12] of the wolf; the whelps of this species, it is supposed, were captured by primitive men and brought under domestication. Savages, like children, are much given to bringing the young of wild animals to their homes; if the conditions are favorable they will care for these captives, even if the charge upon their resources is tolerably heavy. With most primitive people, however, life is so vagarious and starvation so recurrent that they are not apt to retain their pets long enough to establish domesticated forms. Thus, among our American Indians, though they show fondness for wild creatures as much as any other people, no species save the dog ever became permanently associated with their tribe. It is, however, possible, that in some sedentary group of savages the work of domesticating the ancestors of the dog, even if they were wolf-like, was accomplished.

The difficulty of this view is that even with the high measure of care which the conditions of civilization permit us to devote to the effort, it has been found impossible to educate captive wolves to the point where they show any affection for their masters, or are in the least degree useful in the arts of the household or the occupations of the chase. They are, in fact, indomitably fierce and utterly self-regarding. It seems unreasonable to believe that any savage would have found either pleasure or profit from an effort to tame any of the known species of wolves. Moreover, the fact that dogs show little or no tendency to revert to the form and habits of their brutal kindred, or to interbreed with them, is clearly against the supposition that there is any close relation between the creatures.

Greyhound after "the Kill"

Yet other speculative inquirers have sought the origin of [13] the dog through the admixture of the blood of several different species, the wolf and the jackal being, perhaps, the principal or the only components of the hybrid stock. Here, too, the evidence of nature is against the supposition. No one has ever succeeded in hybridizing the wolf and the jackal, nor do our dogs show any more tendency to revert to the jackal than to the wolf. They meet their tropical relative with as much animosity as is proper, or at least customary, in the intercourse of allied yet distinct species. In fact, all the indices by which we are able to carry back the history of other domesticated animals to their primitive or even extinct ancestry, fail in the case of the dog. When the stock is allowed to go as nearly wild as they can be induced to become, we do not find that they thereby approach to any known wild form. It therefore seems [14] reasonable to betake ourselves to another basis for the natural history of the dog, which has not yet been made a matter of much inquiry, but which promises to afford us more substantial truth than the conjectures which we have just considered.

We should, in the first place, note the fact that the ancestors of our more important domesticated animals, those which have been longest in subjugation, have commonly disappeared from the wild state—the species, except for the cultivated forms, having gone into the irrecoverable past. This is the case with the wild kindred of our bulls, horses, sheep, and camels, there probably being none of the original wild species of these groups now living, except those which have been more or less completely subjugated by man, and then have returned to the wilderness. The fact is, that with any large mammal the domestication of the species tends to bring about the destruction of the remaining wild forms. If we go back in fancy to the time when the dog was taken in from the wilderness, we readily perceive how certainly the subjugated individuals would have mingled with their wild kindred, so that either the wild would have become tame or *vice versa*. The same incompatibility which exists between slavery and freedom in our own species in any given territory may be said to hold in the case of captive animals. It is particularly on this account that I am disposed to think that our races of dogs have been derived from one or more original species of truly canine ancestors, the wild forms of which have long since disappeared from the earth.

St. Bernard

Although there are no species of wild dogs now in existence to which we can refer the origin of our household friends, [15] there are several known to us only in their fossil state, from which they may possibly—indeed, we may say probably—have been derived. These creatures are, of course, represented only by their skeletons, and even these remains have only been found in an imperfect state of preservation. It is evident, however, that these extinct species, or at least certain of them, lived down to the time when man had come upon the earth, and was beginning to speculate on his surroundings for such company and help as he might win therefrom. It may interest the reader to know that a species of American dog existed in the Southern Appalachians down to a very recent time—recent, at least, in a geological sense. The remains of one of these animals were found by the writer in a cave in East Tennessee, near Cumberland Gap. From the fragments of the skeleton, Mr. J. A. Allen has described the species. The animal [16] appears to have been of moderate size, and, from the position of the bones, it seems tolerably certain that it lived but a few centuries ago.

It is clearly a reasonable supposition that some of these primitive canine species may have been far more domesticable than the existing kindred of the dog—the wolves, foxes, jackals, or hyenas—

differing from their fiercer kindred much as the zebras do from the wild asses, the one form being utterly undomesticable, and the other lending its back almost willingly to the burdens which man chooses to impose. It seems likely that this primitive species—perhaps more than one—whence the dog sprang was not a very vigorous or widespread form; else, as before remarked, a savage would have found it impossible to keep his half-tamed creatures from rejoining their wild kinsmen. Thus, if a man should in this day succeed in taming wolves, in a region where they were plenty, to the point where they began to abide his presence, or even to have some slight affection for him, the call of nature would be likely to lead them back to reunion with their kind.

It seems pretty certain that the first steps in the domestication of the dog must be attributed not to any distinct purpose of acquiring a useful companion, but to that vague instinct which leads children to make captives of any wild animals with which they come in contact. The fancy for pets is not only common to all mankind, civilized and savage alike, but is clearly exhibited in many of the mammals below the level of man. Almost every one has observed cases where dogs, cats, and horses have become attached to some creature of an alien species with which they have been by chance thrown in contact. The higher the grade of the intelligence, the [17] more sympathetic with other life the animal is likely to become. Thus the elephants, whose natural endowments in the way of intelligence are perhaps superior to those of any other wild creatures, are, when brought into captivity, curiously prone to form attachments to human beings. Savages appear to make but little use of their dogs in hunting. In fact, those peculiar combinations of instinct and training which we find in our hounds, pointers, setters, and other dogs which have been bred to serve the purposes of sportsmen, have been acquired but slowly, and are of no value except where the search for game is carried on under what we may term civilized conditions. The dog of the savage is in all countries much like his master—a creature with few arts and unaccustomed to subdue his rude native impulses.

Spaniel Retrieving Wild Duck

It seems most likely that for ages the principal use of the [18] dog which dwelt about the camps of the primitive people was found in the reserve food supply which they afforded their thriftless masters. When the hunting was successful the poor brutes had a chance to wax fat, and even in times of scarcity they managed to pick up enough food to keep them alive. When their masters were brought to a state of famine they were doubtless accustomed, as are many savages at the present time, to eat a portion of their pack. In the early conditions of humanity there was no other beast which could be made to serve so well this simple need in the way of provender. The dog is, in fact, the only animal ever domesticated which can be trusted through his own affections alone to abide with his master in the endless changes of camp and the rapid movements of flight and chase which characterized men before their housed state began. In a certain curious way the use of dogs for food has served greatly to advance the development of these captives. When the savage was driven to feed upon his dogs he was naturally more willing to sacrifice the least intelligent and affectionate of them, delaying, to the point of extremity, the time when he would kill those which had endeared themselves to him. In this way for ages a careful though unintended process of selection was applied to these creatures, and to it we may fairly attribute, as many considerate naturalists have done, a large part of the intellectual—indeed, we may say moral— elevation to which they have attained.

When the place of the dog as the first and most intimate companion of man was affirmed in the rude way above described—when the savagery to which he was at first made free gradually enlarged to civilization, a number of special uses were found for the peculiar capacities of the creature. [19] These varied in the different parts of the world, according to the peculiarities in the conditions of the masters. In high latitudes, where the ground is snow-covered during the winter season, dogs were used, as they are to this day, in dragging sleds. They were, indeed, perhaps the first animals which were harnessed to vehicles. When they were brought to serve this definite end, we may well believe that the stronger and more enduring individuals were spared in times of dearth for the reason that they were almost indispensable to their masters, and even the little forethought which we find among primitive peoples would lead to their preservation. Here again, doubtless, came in the process of unintended selection which has made the Esquimau sled-dog one of the most remarkable varieties of his kind.

Perhaps the most interesting of the early variations induced among dogs is that which has arisen from the pastoral habit. We do not know when this custom of keeping sheep in large flocks was first instituted, but it is evidently of exceeding antiquity, probably far older than the pyramids of Egypt. The custom could hardly have been instituted without help of the shepherd's mate, the sheep-dog. Although the creatures of this breed are probably in form very near to the original wild species whence our canines came, the variety has as regards its instincts been, by a process of education and selection, led very far away from the original stock.

The wild forefathers of this species were clearly natural born sheep-slayers, and the motive abides to this day in all the breeds which have the strength to assail our unresisting flocks. The spirit is so ingrained that even the most civilized of our house-dogs, which may for generations never have [20] tasted blood and which show no disposition to attack the other animals of the barn-yard, cannot be trusted alone with sheep. When two or more of them are together the old instincts of the wild pack return, and they will slay with insensate brutality until they are fairly exhausted with their fury. Their behavior on such occasions reminds one of the actions of their masters when possessed with the blind rage of a mob. Yet in the shepherd-dog we find this ancestral motive, once a large part of the life of the creature, so overcome by education and selection that they will not only care for a flock with all the devotion which self-interest can lead the master to give to the task, but they will cheerfully undergo almost any measure of privation in order to protect their charges from harm. The annals of shepherd districts, especially those where winter snows fall deeply, as in Scotland, abound in anecdotes of a well-attested nature which show how profoundly the dogs which tend the flocks are imbued with the love of the animals committed to their care. This affection is more curious for the reason that it is never in any measure returned by the sheep. To them the custodian is ever a dreaded overseer. He seems to bring to them nothing but the memories of danger derived from the experience which their species acquired in far-away times.

It is very interesting to note the behavior of a young shepherd-dog when he is first brought in contact with a flock. It is easy to see that he has an amazingly keen interest in the sheep. He regards them with an attention which he gives to no other living things, except perhaps his master. Out of a litter of well-bred pups belonging to this variety, the greater part will at once assume a curatorial attitude toward a flock. They will show a disposition to keep them [21] together, and will seize on an individual only in case he undertakes to break away. They will generally use no more force than is necessary to reduce the recalcitrant to order. They arrest him by catching hold of the leg or fleece, and rarely seize hold of the throat, which other dogs, led by their inherited instincts, are apt at once to assail. Very rarely does a shepherd-dog of good ancestry, even at the outset of his career, attack a sheep in a way which shows that the ancient proclivities have been revived in his spirit. Even then a little remonstrance, or at most a slight castigation, is pretty sure to turn him from his evil ways. If we could measure in some visible manner the psychic peculiarities of animals, we would be led to regard this great change in the instincts of the dog, which has been brought about by his use in herding, as perhaps the most momentous transformation which man has ever accomplished in any creature, including himself; for none of our own inherited savage traits are so completely sublated at the time of our birth as is this old and sometime dominant slaying motive in the shepherd-dog.

With the advancing differentiation of human occupations and amusements, our breeds of dogs have, by more or less deliberate selection, been developed until by form and instincts they fit a great variety of purposes. Some of these pertain to industrial work, but the greater portion are related to the sports or fancies of men. The turnspit was bred for its short legs and small, compact body, and was serviceable in those treadmills of the hearth which have long since passed out of use, but which were for centuries features in our kitchens.

Bull-Dog

The massive type of bull-dogs, characterized by heavy frames and an indomitable will, appears to have been brought [22] about by a process of selection having for its unconscious end the development of a breed which should render the herdsman of horned cattle something like the assistance which the shepherd-dog gave to those who had charge of flocks. In the more primitive state of our bulls and cows the creatures were much wilder than at present, and were generally kept, not in enclosed pastures, but on unfenced ranges. In these conditions the care taken needed the help which the ancestors of our modern bull-dog afforded. The tasks which the animal was called on to perform were of a ruder nature than those which were allotted to the shepherd-dog. Their business was to conquer the unruly beast. They were taught to seize the muzzle, and by the pain they thus inflicted they could subdue even the fiercer small bulls of the ancient type of form. From this original use the cattle-dogs were turned to the brutal sport of bull-baiting, a rude diversion which was indulged in by our ancestors for centuries, and has only dis [23] appeared in our less cruel modern days. Bred for the bull-ring, these dogs acquired the formidable strength and ferocity under excitement which made their name a terror and their qualities a satirical embodiment of the ruder traits which characterized the British folk.

The training which instituted the breed of bull-dogs was evidently much less continuous and effective than that which developed the shepherding variety. The use for the creature in the care of herds has passed away. In the older parts of the world cattle are kept only in enclosures; and where, as on our frontier, they still range over unbounded fields, they are guarded by horsemen who do not need the assistance of dogs to control the movements of the herds. No longer serviceable either in economies or sports, the breed of true bull-dogs is rapidly disappearing. As we may often observe in other fields of development, the peculiarities of this breed are now under the control of fancy, and the blood is being led far away from its old characteristics. The bull-terrier and other varieties, which retain something of the form and of the solemn demeanor which characterized their ancestors, but which are too small to assail horned cattle, mark the vanishing stages of this great stock, which will soon be known only in memory. The history of this peculiar herd-dog shows us how marvellously pliant the body and mind of this species has become under the conditions of civilization. The rude process of unconscious selection, acting without steadfastness of purpose or rationally developed skill, serves to sway the qualities of the animal this way or that to meet the ever-changing requirements of use or fancy. A similar selection in the case of our horned cattle has within a few centuries converted the cows into mild-mannered and [24] sedentary milk-making machines, and has deprived the bulls of the greater part of their ancient savage humor. Owing to this change in the quality of their associates in captivity the dogs have also been led into great variations. The same type of interaction may be traced again and again in the isolated part of the world enclosed within our fences, as well as in the free realm of the wildernesses. All the individuals in the great host of life affect each other as do the soldiers of a well-organized army in the movements of a battle.

The shepherd-dog, the turnspit, and the bull-dog are the three remarkable variations of the canine blood which were brought about by a process of training and selection unconsciously directed to the institution of breeds suited to special economic ends. The other varieties of dogs have been shaped more distinctly for purposes of amusement or for the indulgence of mere fancy. The several varieties of hounds, harriers, beagles, pointers, setters, terriers, etc., have been designed to meet a dozen or more variations in the conditions of the chase. The marvellously complete way in which special peculiarities have been developed in mind and body makes this field of domestic culture the most fascinating subject of inquiry to the naturalist. The ordinary fox-hound has had his inheritances determined so as to fit him for pursuing a small animal which can rarely be kept in view during its flight, and which can only be followed by the odor it leaves in its trail, so these creatures run almost altogether under guidance of their sense of smell. The stag-hound, on the other hand, pursues a relatively large animal which cannot well be followed by the nose, at least with any speed; they therefore trust almost altogether to vision in their chase. The packs which hunt otters have developed the swimming [25] habit and an array of instincts which fit them especially for this peculiar sport. If space allowed we could note at least a dozen divisions of the group of hounds or chasing dogs, each of which has developed a peculiar assemblage of qualities, more or less precisely adapted to some particular game.

Fox-Hound and Pups

Perhaps the most special adaption which man has brought about in his domesticated animals is found in our pointers and setters. In these groups the dogs have been taught, in somewhat diverse ways, to indicate the presence of birds to the gunner. Although the modes of action of these two breeds are closely related, they are sufficiently distinct to meet certain differences of circumstances. The peculiarities of their actions, it should be noted, are altogether related to the qualities of our fowling-pieces. These have been in use, at least in the form where shot took the place of the single ball, for less than two centuries, and the peculiar training of [26] our pointers and setters has been brought about in even less time. It seems likely, indeed, that it is the result of about a hundred and fifty years of teaching, combined with the selection which so effectively works upon all our domesticated creatures. It thus appears that this peculiar impress upon the habits of the hunting-dog is the result of somewhere near thirty generations of culture.

Pointer Retrieving a Fallen Bird

Although, as has been often suggested, the pointing or setting habit probably rests upon an original custom of pausing for a moment before leaping upon their prey, which was possibly characteristic of the wild dog, it seems to me unlikely that this is the case, for we do not find this habit of creeping on the prey among our more primitive forms of dogs nor the wild allied species as a marked feature. All the [27] canine animals trust rather to furious chase than to the cautious form of assault by stealthy approach and a final spring upon their prey, as is the habit with the cat tribe. Granting this somewhat doubtful claim that the induced habits of these dogs which have been specially adapted to the fowling-piece rest upon an original and native instinct, the amount of specialization which has been attained in about thirty generations of care remains a very surprising feature, and affords one of the most instructive lessons as to the possibilities of animal culture.

Pointer and Setter, Flushing Game

It is an interesting fact that the variation of a spontaneous sort, which is now taking place in our pointers and setters, is considerable. It is, perhaps, more distinctly indicated here than in any other of the breeds which are characterized by peculiar qualities of mind. All those familiar with the behav [28] ior of these strains of dogs have observed the high measure of individuality which characterizes them. I have recently been informed by a friend, who is a hunter and a very observing naturalist, of one of these variations in the pointer's instinct, which may, by careful selection, possibly lead to a very useful change in the habits of the animal. Hunting the Virginia partridge in the tall grass on the sea-coast of Georgia, his dog found by experience that his master could not discern him when he was pointing birds, and that a yelp of impatience would put up the covey before the gun was ready for them. The sagacious dog, therefore, adopted the habit of backing away from the point where he first fixed himself, so that he, by barking, denoted the presence of the birds without giving them alarm. Although, in this first instance, the action is purely rational, and is indeed good evidence of singular discernment and contriving skill, it seems likely that by careful breeding it may be brought into the realm of pure instinct or inherited habit.

The great variation in habits which is taking place in those varieties of dogs which are immediately under the master's eye during all the process of the chase, is easily explained by the fact that these creatures are in a position to be immediately and constantly influenced during their most active, and therefore teachable state of mind, by the will of man. A pack of fox-hounds is, to a great extent, out of hand while engaged in the pursuit of their prey; but a pointer or setter, even when under extreme excitement, is almost completely mastered by the superior will. When we observe the extent to which human intelligence is affecting the qualities of our hunting-dogs, it is not surprising to note that, in almost every district where there are peculiar kinds of game, varieties [29] of the dog are developing which are especially adapted to its pursuit. Thus, in the parts of North America where the raccoon abounds, a variety of hunting-dog is in process of development which has a singular assemblage of qualities which fit it for this peculiar form of the chase. Although as yet "coon-dogs" have not been cultivated for a sufficient time to acquire distinct physical characteristics, their habits exhibit a larger range of specialization than those of any other breed of sporting dogs.

In those parts of the Americas where peccaries are hunted, the dogs used in their pursuit have learned to beware of assaulting the pack which they have brought to bay, and instead of indulging in the instinct which leads them into that way of danger and of certain death, they circle round the assemblage, compelling them to show front on every side and so to remain stationary until the hunters come up. Perhaps a score of similar specializations in the modes of action of our dogs which are employed in the chase could be recited; but as they all lead us to one conclusion—which is to the effect that these creatures are, as far as their mental powers are concerned, like clay in the hands of the potter—we may pass them by for some considerations which appear to have escaped the attention of writers who have discussed the problems of canine intelligence.

The singular elasticity as regards both mental and physical qualities which the dog exhibits, may well be compared with the other conditions which we find in certain of our domesticated animals, as, for instance, in the horse, where the mind shows but slight changes, and where the body has proved far less plastic than among dogs. The readiness with which the proportions of the dog may, by the breeder's [30] art, be made to vary, is probably due to the fact that the group to which this creature belongs is one of relatively modern institution. It has the plasticity which we note as a characteristic of many other newly-established forms. The flexibility of mind is a concomitant of the carnivorous habit where creatures obtain their prey by the chase. Such an occupation tends to develop agile minds as well as bodies, and where exercised as it doubtless was by the ancestry of the dog, in the manner of pack hunting, where many individuals share in the chase, it is well calculated to insure a certain free and outgoing quality of the mind.

Dutch Dogs used in Harness

So long as our dogs were employed in the labor or the organized recreations of man, the tendency of the association with the superior being was in a high measure educative. They were constantly submitted to a more or less critical [31] but always effective selection which tended ever to develop a higher grade of intelligence. With the advance in the organization of society the dog is losing something of his utility, even in the way of sport. He is fast becoming a mere idle favorite, prized for unimportant peculiarities of form. The effort in the main is not now to make creatures which can help in the employments of man, but to breed for show alone, demanding no more intelligence than is necessary to make the animal a well-behaved denizen of a house. The result is the institution of a wonderful variety in the size, shape, and special peculiarities of different breeds with what appears to be a concomitant loss in their intelligence. We often hear it remarked by those who are familiar with dogs that the ordinary mongrels are more intelligent and more susceptible of high training than the carefully inbred varieties, which are more highly prized because they conform to some thoroughly artificial standard of form or coloring. This is what we should expect from all we know concerning the breeding. Where for generations the dog-fancier has selected for reproduction with reference to the trifling and often injurious features of shape he seeks to attain, he naturally and almost necessarily neglects to choose the creatures in regard to their mental peculiarities. The result is that the breed tends to fall back in these regards to below the level of the ordinary cur, who makes his place in the affections of his owner because he has attractive or useful qualities of mind. It appears to me, in a word, that our treatment of this noble animal, where he is bred for ornament, is in effect degrading.

Although the formation of our fancy breeds does not serve to advance the development of those intellectual feat [32] ures which are the most interesting part of our dogs, the experiments have served to show the amazing physical plasticity of this species under the conditions of long domestication. The range in size between a tiny spaniel, such as those which are bred in Chihuahua, in northern Mexico, and the great Danes or mastiffs of northern Europe, is, perhaps, the greatest which has ever been attained in any mammal. In some cases the larger individuals belonging to the mastiff breed probably weigh nearly thirty times as much as their smaller kinsmen. Great as are these variations, they are only in form and bulk. They involve none of those curious changes in the number of bones of the skeleton which we may trace among the domesticated pigeons. We therefore turn from these results of breeders' fancy to consider certain of the mental qualities of dogs which have not come in our way in our review of the history of its relations to man.

King Charles Spaniel

First of all, we may note the fact that the friendly relations which dogs have become accustomed to form with men vary exceedingly in their range and activity. Perhaps in no other regard does the dog exhibit such distinctly human characteristics as in the way in which he meets the individuals of the mastering species. The gamut of their social relations with men is almost exactly parallel with our own. With from one to a dozen persons a dog may maintain an attitude of almost equally complete sympathy and mutual understanding. He may be on terms of acquaintanceship in varied degrees of familiarity with a few score others with whom he comes in frequent contact. Toward the rest of mankind he maintains a position of more or less complete distrust, which with experience may attain the indifference which men commonly [33] show toward perfect strangers. If we observe a dog going along a much-frequented street, we may note that his relations to the people are substantially those which the folk have to each other. He shows as they do a certain consideration for the individuals he encounters, gives them their due place, and yet holds to his own. It is particularly noticeable that he avoids all contact with the other passers—in fact a dog has to be much beside himself with rage or fear, or insane from disease, before he will break those bounds of personality which civilization has set up to guide the conduct of life.

The social culture of dogs appears to have gone to the point where they recognize the meaning of an introduction—at least as far as the sympathetic relations of that understanding are concerned. Almost any well-bred dog will submit to be presented by his master, or even by persons whom he knows but is not accustomed to obey, to a stranger to whom he has already exhibited some dislike. During the introduction he will submit to those formal exchanges of courtesy which he is accustomed to recognize as the indices of friendship. The impression of this understanding seems to be so permanent that on subsequent meetings the dog, [34] though he may maintain his original dislike of the man who has been forced upon his acquaintance, will continue to treat him with a certain consideration, though it is often easy to see that it is a difficult matter for him to conform to the requirements of society. When we compare the conduct of dogs in these regards with the behavior of other animals, even highly domesticated forms, we perceive how marvellously successful has been man's unconscious effort to mould this creature on his own nature.

Another extremely human characteristic of our canine friends is shown in their susceptibility to ridicule. Faint traces of this quality are to be found in monkeys and perhaps even in the more intelligent horses, but nowhere else save in man, and hardly there, except in the more sensitive natures, do we find contempt, expressed in laughter of the kind which conveys that emotion, so keenly and painfully appreciated. With those dogs which are endowed with a large human quality, such as our various breeds of hounds, it is possible by laughing in their faces not only to quell their rage, but to drive them to a distance. They seem in a way to be put to shame and at the same time hopelessly puzzled as to the nature of their predicament. In this connection we may note the very human feature that after you have cowed a dog by insistent laughter you can never hope to make friends with him. A case of this kind is fresh in my experience. A year or two ago I was imprudent enough to laugh at a very intelligent dog in my neighborhood, he having unreasonably assailed me at my house-door, where he had been left for a long time to wait while his owner was within and had thereby been brought into an unhappy state of mind. Sympathizing with his situation, I preferred to laugh him out of his humor [35] rather than to beat him with my stick. I regret I did not take the other alternative, for I made the poor brute my implacable enemy by my pretence of contempt for him. I am inclined to think that if I had beaten him the matter could have been arranged afterward in a friendly way.

The Pounce of a Terrier

Another very remarkable and I believe hitherto unnoticed likeness between the mind of dogs and that of man is found in the fact that these dumb beasts, unlike all other inferior animals, except, perhaps, some of the more intelligent species of monkeys, will learn lessons from isolated experiences. In this regard they are indeed quite as apt as the lower kinds of men. Thus a dog who has had an unsavory or painful experience with a skunk or a porcupine is apt to keep away [36] from these creatures for a long time thereafter. Where, as is not infrequently the case, a cur takes to eating eggs, a single dose of tartar emetic concealed in an egg which is placed where he can readily find it, is apt to effect an immediate and complete reform. This ready learning from experience is almost the gist of our human quality—at least on the intellectual side of it.

Perhaps the greatest success to which man has attained in his education of the dog is to be found in the measure in which he has overcome the fierce rage which clearly characterized the ancestors of this creature when they first felt the mastering hand. The reader cannot understand the intensity of the rage motive in the carnivora unless he has studied some of these brutes in their wild state, where from the time in the remote ages when they first began to take on the qualities of their species they have survived and won success by the fury of their assault. In almost all our breeds of dogs this primal ferocity has been overlaid by the various motives of rationality, sympathy, and conventional demeanor, until one may live half a lifetime with well-bred dogs without a chance to see the demon which we have buried in their breasts, as we have in our own, beneath a host of civilizing influences. It is rare indeed in our day that a dog, unless insane, will bite a human being. The most of their assaults are pure bluster, mere pretence of fury, as is shown by the fact that if, carried away by their pretence, they are led to use their teeth, it is usually a mere sham assault, having no semblance of the effectiveness of true combat.

Something of the pristine fury of the primitive dogs may still be noted in a certain brutal variety of watch-dogs which are still to be found in parts of continental Europe. The [37] best types of this breed which I have ever seen are to be found among the dogs which are kept to guard the quarries of Solenhofen, in Bavaria, whence come all the fine lithographic stones which are so extensively used in printing. These quarries are scattered over several square miles of untilled country, and the separate pits are to be numbered by the score. As much valuable stone is necessarily left over night in the quarries, their care is confined to packs of watch-dogs which are turned loose at night and appear as if by instinct to spend the hours of darkness in prowling over the territory. Such is their size and ferocity that it takes a sturdy beggar to face them. I remember inadvertently disturbing one of these brutes from sleep, in the strong cage where he was confined, and I have never beheld such a picture of blind fury as he exhibited. I had not come within twenty feet of him, and was merely moving past his place of confinement; yet he sprang to the grating and strove with his teeth to break his way through the bars. I thought the animal must be mad, but his keeper assured me that such was his ordinary state of mind and that the humor was common to all the breed; even the masters dwelt in fear of them. Ordinarily the only exhibitions of the innate ferocity of our dogs are to be seen in their combats with each other, when for a time the creatures return to their primitive state of mind. Even these occasional exhibitions of fury are not found among all breeds of dogs, and among many individuals even of the combative strains of blood the motive of battle appears to have quite passed away.

Pomeranian or "Spitz"

In antithesis to the old Ishmaelitic humor of our primitive dogs, man has developed a singular, sympathetic, and kindly motive in these creatures. From the point of view of [38] the dog's education we must not set too much store by his affection for his master. This kind of devotion of one being to another is displayed elsewhere in the animal kingdom, though it is more common among birds than among mammals. We find traces of it in the greater part of our domesticated creatures or in those which we have individually adopted from the wilderness. It is a part of the great sympathetic motive, which, originating far down in the series of animals, increases as they gain in the scale of being, until it reaches the highest level it has yet attained in spiritually minded men. The eminent peculiarity in the case of a dog is that the very centre of his life is formed of the affections, which are evidently the same as those which rule the days of the most cultivated men. To him these elements of friendliness are absolutely necessary to a comfortable existence. If by chance he becomes separated from his master and the other people with whom he is familiar, his bereavement is intense; but in most cases, at the end of a day or two, he is compelled to form new bonds, and he sets about the task in an exceedingly human way. I dwell in a town where dogs abound and where the frequent coming and going of the people puts many of the creatures astray. Perhaps as often as once a week, almost always late in the evening, one of these [39] unhappy lost ones seeks to make friends with me. His advances toward this end always begin by his dogging my footsteps at a little distance. If I do not repulse him he will come nearer until he has made sure of my attention. A friendly word will bring him to my hand; but his behavior is never effusive, as it would be if he had found his rightful owner, but mildly propitiative and with a touch of sadness. There is, it seems to me, no other feature in the life of the dog which tells so much as to his moral nature as his conduct under these unhappy circumstances.

Poodles

In the long catalogue of human qualities which characterize our thoroughly domesticated dogs, we must not fail to take account of their sense of property. In this the creature differs from all other of our domesticated animals. It is a common characteristic of mammals, both in their wild and tame state, that they feel a motive of ownership in the food which they have captured or in the den which they have made their lair; but beyond these narrow personal limits we see no evidence of any sense of ownership in land or effects. We readily observe, however, that our household dogs not only know the chattels of their master and distinguish them [40] from those of other people, but they also learn to recognize the bounds of their house-lot or even of a considerable farm. When a dog, even of a militant quality, enters on territory which he does not feel to belong to him, he is at once a very different creature as compared to his condition when he is on his own land. He treads warily and will accept without dispute an order to take himself off. A perception of this sort indicates an extraordinary amount of sympathy and discernment. It requires us to assume that the creature has a good sense of topography and that he observes closely the various acts, none of them perhaps very indicative, which go to show the limits of his master's claims.

Although the mental qualities of our highly domesticated dogs are singularly like those of their masters, the likeness going to the point that the household pet is apt to have acquired something of the general character of the people with whom he dwells, there are many suggestive differences arising from failures of development which are in the highest measure interesting to those who study the species. We note, in the first place, that although for ages in contact with the constructive work which occupies his masters, the dog shows no tendency whatever to essay any undertakings of this nature. He is quite alive to considerations of personal comfort and is particularly fond of a warm bed; yet, except for a few unverified stories, we may say that there is no evidence whatever to show that they ever try to improve their conditions by deliberately providing themselves with warm bedding. In no well-attested case has a dog shown any sense as to the nature of any mechanical contrivance. They will learn which way a door opens, and rarely if ever do they undiscerningly close it when it is slightly ajar and they [41] wish to pass through the opening; but I have never been able to observe or obtain evidence to show that they would without teaching pull down a latch in the way in which a cat readily learns to do. Much as dogs have had to do with guns, they display no kind of interest in the arms except so far as they are tokens of sport to come. They connect the explosion with the capture of game, and will search for it in the direction toward which the barrel was pointed. I have not, however, been able to find that they know, as they might readily do, and as a crow would surely do, when the weapon was loaded and when empty. They show no interest in it, such as monkeys readily display toward any mechanical contrivance to which their attention has been directed. All these negative features indicate that the mechanical side of the canine mind is entirely undeveloped.

Collie

[42]

Although there is some evidence that the sense of number attains a measure of development in dogs, the ability to form mathematical conceptions of any kind appears to be very weak in this species. The fact that shepherd-dogs, in a way, keep an account of considerable flocks so that they will know when one is gone astray, can readily be explained on the supposition that they know their charges individually and not in sum. The absence of arithmetical capacity is, however, less important than the lack of mechanical sense, for the reason that such incapacity is also common in the lowest races of men. Although dogs, as before noted, quickly and clearly acquire a notion of property rights in all which pertains to their owner's holdings, they appear never to extend their sense of their own personal possessions beyond the original limit to which they had attained when the species was domesticated. The creature feels a sense of personal property in his food and in his sleeping-place, but appears not to extend his conception of individual rights beyond these primitively established limits.

All our well-bred household dogs quickly learn certain bodily habits which are necessary to make them acceptable members of a household. These habits are not well affirmed by inherited instinct, but the ease with which the instruction is acquired shows that they have become prone to submit to such regulations. Culture on this line rests upon a primal instinct, originating we know not how, which leads a number of wild animals to conceal their excrement. On the other hand, these creatures exhibit no sense of modesty, though that, in a more or less complete measure, is characteristic of all human tribes whatsoever.

As regards the memory, dogs appear to have a considerably [43] greater measure of capacity than is observable in any other group of domesticated animals. There is no question that they can recall their associations with people from whom they have been separated for a year or more. Some trustworthy anecdotes appear to establish the fact that the recollections may endure for two or three years. I have observed an instance in which the memory seems perfectly clear after an interval of eighteen months, and this concerned a person who had been with the dog for a period of not more than four days. It is interesting to note the behavior of a dog when he has failed to recognize a person whom he has known well, but from whom he has been long separated. I have a shepherd-dog that has known me well, but the friendship is often interrupted by partings of some months' duration. When, after one of these absences, I appear to him in the distance, he comes furiously towards me, quite possessed by his enmity. At a certain point in his charge a doubt begins to beset him; he moderates his pace; his roaring bark passes into a whine; and as the full measure of his blunder is borne in upon him by my voice, he becomes the picture of shame. In his perplexity, he always finds relief in endeavoring with his paw to scrape a supposititious fly from the side of his nose. He then deals with what I suppose to be an equally imaginary flea; after he has thus gained a few seconds for readjustment, he welcomes me joyously. All this is so thoroughly human-like, that even the naturalist, the professional doubter, is forced to believe that the dog's mind works substantially as his own, and that the feelings connected with the action are essentially the same.

While in the case of the elephant and the pig, and in a less measure in several other of the lower animals, we have [44] indices of as high or even higher intelligence than the dog, no other brute shows anything like the same measure of what we may term human quality. So far as the field of the emotions is concerned, we are driven to believe that it has been bred into the kind by the ages of intimate associations, supported by the selective process which has led people to preserve the individual of the species with which they found themselves the most in sympathy. I repeat the suggestion, and shall repeat it yet again, for the reason that just here—how effectively the reader's imagination will suggest—we find a basis for the hope that, with time and care, man may bring his subjects of the lower realm into a more intimate, affectionate, and helpful relation than is dreamed of by those who look upon them as mere brutes.

The most curious limitation which we find in dogs is as to the measure of expression to which they have attained. No one who has well considered the facts can doubt that our civilized varieties of this species have something like a hundred times as much which deserves utterance as their savage forefathers possessed. Yet the capacity for giving note to these thoughts or emotions has not gained anything like the proportion to the needs. It seems, however, that some gain in this direction has been made, and that much may be won hereafter in the way of further advance. Never having known the species whence our dogs came in its wild state, we are uncertain as to its modes of expression; but, observing the varieties of dogs which are kept by savages, it seems probable that the primitive canines used their voice only in howling or yelping; that is, as a continuous sound akin to the bellowings or other cries of the various wild mammals. It is characteristic of all these primitive forms of utterance [45] that they are, to a great extent, involuntary, and that when the outcry is begun it continues in a mechanical manner, with no trace of modulation arising from the conditions of the moment. In other words, these actions resemble, in a way, sneezing or hiccoughing in human kind; actions which are stimulated by certain states of the body, but which are not at all under the control of the will. Howling or bellowing doubtless represents, in a measure, a state of mind as well as of body, but the action is of a general and uncontrolled kind.

The effect of advancing culture upon a dog has been gradually to decrease this ancient undifferentiated mode of expression afforded by howling and yelping, and to replace it by the much more speech-like bark. There is some doubt whether the dogs possessed by savages have the power of uttering the sharp, specialized note which is so characteristic of the civilized forms of their species. It is clear, however, that if they have the capacity of thus expressing themselves, they use it but rarely. On the other hand, our high-bred dogs have, to a great extent, lost the habit of expressing themselves in the ancient way. Many of our breeds appear to have become incapable of ululating. There is no doubt but this change in the mode of expression greatly increases the capacity of our dogs to set forth their states of mind. If we watch a high-bred dog, one with a wide range of sensibilities, which we may find in breeds which have long been closely associated with man, we may readily note five or six varieties of sound in the bark, each of which is clearly related to a certain state of mind. The bark of welcome, of fear, of rage, of doubt, and of pure fun, are almost always perfectly distinct to the educated ear, and this although the observer may not be acquainted with the creature; if he knows him well, he may [46] be able to distinguish various other intonations—those which express impatience and even an element of sorrow. This last note verges toward the howl.

It does not seem to me that we should regard barking as a new and useful invention; there are, indeed, few such in the organic world. The sound appears to me to have been derived from the primitive habit of howling. If we hearken to this utterance we perceive that it is not an unbroken sound, but is somewhat intermittent. At either end of the prolonged sound we can often notice that it is divided into rather distinct yelps more or less completely separated from the other notes. The cries of a dog when beaten often exhibit the same peculiarity; so, too, the puppy, before he has attained skill in barking, will often prolong each utterance in a way which makes its relation to the ancient mode of expression tolerably clear. At the risk of being deemed fanciful, I venture to suggest that the bark is in effect a division of the howl into clearly separated notes, the change having come about as a similar alteration is effected in our own speech, by the increase in the intelligence which the creature is called upon to express. I conceive that while the primitive and massive emotions found satisfying utterance in the long-drawn notes, the more divided state of mind of the humanized successor has led to a change in its utterances. Although these modifications of speech, if such we may term them, have probably been developed on the basis of the dog's human relations, there is, it seems to me, good reason to believe that the diversities in note have come to have a distinct conventional value between the individuals of all the different breeds. Any one who closely observes these animals must have noticed the fact that the degree of attention they [47] give to the utterances of their kindred varies in a way which indicates that they have great varieties of denotations. Some of the shades of the meaning which a dog's bark has to others of his species probably escape our less fine ears.

The creation of something like a language among our civilized dogs has naturally been accompanied by the development of an understanding of human speech. Although we cannot attach much importance to the mass of anecdote on this point, there is enough which is well attested — sufficient, indeed, which has come within the limits of my own observation — to make it clear that dogs, even without deliberate teaching, frequently acquire a tolerably clear understanding of a number of words and even of short phrases. They will catch these not only when given in distinct command, but when uttered in an ordinary tone, without any sign that they relate to their affairs. It is true that these understood words generally relate to some action which the dog is accustomed to perform, yet there are instances so well attested that they deserve credit, which seem to show that the creatures can get some sense of the drift of conversation even when it is carried on by persons with whom they are not familiar and does not clearly relate to their own affairs.

It should be observed that within the narrow limits of this essay little or no effort has been made to interpret the state of mind of dogs from the vast but rather untrustworthy mass of anecdote with which our books are filled. So large a part of this evidence is contaminated by prepossessions, and a yet larger part is so unverified in any scientific sense, that for purposes of sound inquiry it is worthless. It therefore seems best to limit ourselves, as has been done in this paper, to those general actions of the creatures which are matters [48] of common knowledge and safely beyond question. From these indices we are able to determine a basis for some important conclusions. These are in effect as follows, viz.: Our domestic dog is derived from a species, one or more, akin to the wolf, the jackal, and the fox; to a group of animals not characterized by great native intelligence, but distinguished for their ferocity and their general untamableness. There is no reason to believe that the primitive dog had any more foundation for his great attainments than his obstinately savage kindred, except that he may have had a greater disposition to form an attachment to a master. We can hardly believe that he had any share of that marvellous sympathy with man and understanding of his motives which characterize the high-bred varieties of his species. All this vast transformation, which from a psychological point of view has carried the dog relatively as far up above his origin as civilization has lifted man above his lowest estate, has been due to human intercourse and the long and effective concomitant selection of good from bad. It is hardly too much to say that a large part of our human nature has been transferred into the descendants of this ancient wild beast. The sense of property, a great part of human affections, many of the attributes which constitute the gentleman, have been passed over to him.

In considering the effects arising from the intercourse of man with the dog, we should not overlook the development of human sympathy which has come about through this relation. The fact that the dog has been made by far the most sympathetic of the lower animals, is due to the affection which men for thousands of years have given to him. In his intercourse with this creature, man first learned to develop his [49] altruistic motives beyond the limits of his own kind. With this extension of his affection must have begun the growth of that large motive, which is the most distinguishing feature of our modern life, which leads us to go forth in a loving manner to the living beings about us, not only to our flocks and herds but to the life of the unsubjugated realm as well. Thus, in a way, we may look upon the dog as affording the first steps on the path of culture which was to lift man from his primitive selfishness to the altruistic state to which he has attained.

Great as has been the work of man upon the dog—it deserves, indeed, to be ranked high among all the accomplishments of his culture—there is reason to believe that if he but go forward with understanding in the ways which have hitherto led him blindly to his success, the final result may be very much more perfect than that which has been attained. It is on this account that I feel it fit to make a strong protest against the system our breeders pursue. Except in the case of dogs used in sport and for herding sheep, the sole effort appears to be to create breeds which shall exhibit peculiarities of form which are mere extravagances, and move the real lover of this noble animal to indignation. In these preposterous and unseemly tasks no care is taken to continue the mental development on lines which have been established by long use. Still less is there any effort to essay the development of the intelligence in ways which are clearly open to us, and which afford possibilities of lifting this species to a yet nobler companionship with our own kind.

It seems worth while for our associations of dog fanciers to undertake to develop varieties of dogs solely with reference to the intellectual qualities of the animal. I venture to [50] suggest that those who seek this end should select some of the primitive types of form, such as are found among the undifferentiated mass of the species, those which are improperly termed mongrels, and this for the reason that among these unselected creatures the intelligence is quicker and more varied than it is in the highly developed varieties. Under skilful trainers the successive generations bred in the experimental station should be subjected to tests which will indicate the measure of intellectual ability. The results already attained by the unconscious selection which man has applied serve to indicate that at the end of a century, and perhaps in much less time, we might develop an animal which in various ways would come to a closer intellectual relation with man than any other lower species has attained.

Cats deserve some mention for the reason, that, while they are the least essential, and on the whole the least interesting, of domesticated animals, they have had a certain place in civilization. They afford, moreover, a capital foil by which to set off the virtues of the dog. Nowhere else, indeed, among the creatures which are intimately associated with men, do we find two related forms which afford, along with a certain likeness, such great diversities of quality.

We know nothing as to the time when the cat first found its way to the associations of man. Presumably this period was much later than the advent of the dog into the human family. The presumption rests upon the fact that while the dog does not demand fixed residence as a condition of its fealty, but is at home wherever his master is, the cat is the creature of the domicile, caring more indeed for its dwelling-place than it ever does for the inmates thereof. In a word, the creature must have come to us after our forefathers gave [51] up the nomadic life. Nevertheless, the association is very ancient; it has endured in Egypt at least for a term of several thousand years.

Among the curious features connected with the association of the cat with man, we may note that it is the only animal which has been tolerated, esteemed, and at times worshipped, without having a single distinctly valuable quality. It is, in a small way, serviceable in keeping down the excessive development of small rodents, which from the beginning have been the self-invited guests of man. As it is in a certain indifferent way sympathetic, and by its caresses appears to indicate affection, it has awakened a measure of sympathy which it hardly deserves. I have been unable to find any authentic instances which go to show the existence in cats of any real love for their masters.

In the matter of intelligence cats appear to rank almost as high as dogs. They are even quicker than their canine relatives in discerning the nature of man's artful contrivances; they readily acquire the habit of opening doors which are closed by means of a latch, even where it is necessary to combine the strong pull on the handle with the push that completes the operation. Feats of this sort are rarely if ever performed by dogs.

The most peculiar quality in the mind of cats is the intense way in which they cling to a well-known locality. Their memory of places, and affection for them, if we may so term it, is evidently far greater than that which they feel for people. Some years ago I had an interesting exhibition of this singular humor. A well-grown and thoroughly domesticated cat, one that seemed more than usually attached to people, was brought from my house in town to a place on the [52] shore. When released, the creature seemed for some days to be nearly insane. It did not recognize any of its friends, it betook itself to the fields, and was with difficulty captured at the end of a week of roaming, during which it appeared to have had no food. Confined within one room, it gradually recovered its powers of mind, and began to take account of its friends. In the course of a month it seemed to be reconciled to its surroundings. Nine months after its first sojourn in the wilderness it was again brought from the town to the same place. On the second visit the creature was somewhat uneasy, but this passed away in a day or two. On a third visit, after a like interval, it seemed at once and entirely at home. Nevertheless, its habits while in the country differ very much from those it has in town. In its original domicile it insists on being about the table at meal-times. While in the country it does not care to be present; in fact, it appears to avoid associations with the household. It seems to me that this cat, after the manner of some men whose brains are diseased, now lives in two distinct states of consciousness, each relating to one of its places of abode.

[53-4]

Hounds Running a Wild Boar
(Showing the habit of attacking neck of prey.)

The differences as regards affection for localities which is shown by cats and dogs are perhaps to be accounted for by an original and essential variation in the habits of life in their wild ancestors. Judging by the kindred of the species which are known to us in their wild state, we may fairly suppose that the dogs were of old accustomed to range over a wide field, having no fixed place of abode; the pack ranging, if the occasion served, for hundreds of miles in any direction. On the other hand, with the cats, it is characteristic of the species that they have lairs to which they resort, and a definite hunting ground in which they seek their food. They are, in a [55] word, animals of very determined routine. As there has been no effort by breeding to change this feature, it has remained in all its old ingrained intensity.

As a consequence of the affection which cats have for particular places, they often return to the wilderness when by chance the homes in which they have been reared are abandoned. Thus in New England, in those sections of the district where many farmsteads have of late years been deserted, the cats have remained about their ancient haunts and have become entirely wild. In this State they are bred in such numbers that their presence is now a serious menace to the birds and other weaker creatures of the country. The behavior of these feralized animals differs somewhat from that of creatures which have never been tamed. They have not the same immediate fear of a man, but the least effort to approach them leads to their hasty flight.

While considering the inelastic quality which is exhibited by cats as compared with the dog, the naturalist notes with interest the fact that the former creature belongs to a family which has never been accustomed to any social life beyond the limits of the family. Moreover, all the cats have the habit of hunting in a solitary way, each for itself, in the achievement and in the result. It is otherwise with dogs. They belong to a group which hunts in packs. For ages they have been used to a communal life. Their minds have thus become accustomed to social intercourse; they are used to having their excitements of the chase in comradeship, and generally they are accustomed to the rough-and-tumble fraternity which we behold in a pack of wolves. It was long ago remarked that the really social animals are those which afford the only good material for subjugation. [56] The difference between the cat and dog seems, in a way, to warrant this statement.

Although it is likely that many efforts have been made to domesticate the other larger felines, no distinct success has attended these experiments. A large Asiatic cat known as the chetah is somewhat used in hunting for sport, but the species has never been adopted in any definite way. In fact, with all the larger cats, including the lion, which is structurally a little apart from the other members of the group, the size and furious nature of the animal have made it impossible to begin the process of selection which has been the means whereby the wilderness motive has been replaced by that of the household in the case of all other domesticated beasts.

[57]

THE HORSE

Value of the Strength of the Horse to Man.—Origin of the the Solid Hoof.—Domestication of the Horse.—How begun.—Use as a Pack Animal.—For War.—Peculiar Advantages of the Animal for Use of Men.—Mental Peculiarities.—Variability of Body.—Spontaneous Variations due to Climate.—Variations of Breeds.—Effect of the Invention of Horseshoes.—Donkeys and Mules compared with Horse.—Especial Value of these Animals.—Diminishing Value of Horses in Modern Civilization.—Continued Need of their Service in War.

The largest economic problem which primitive people on their way upward towards civilization had unconsciously to face was that of obtaining some kind of strength which could be added to the power of their own weak limbs. For all his eminent capacities of body, man is not a strong animal, nor is he so built that he can apply the measure of strength that is in him to good advantage. There are scores if not hundreds of species with which he came in contact in his effort to dominate nature that are stronger, swifter, and better provided with natural weapons. With the first step upward, as in almost all the succeeding steps, the advance depended on securing more energy than that with which our kind was directly endowed. It is hardly too much to say that the progress of mankind beyond the savage state would probably never have been effected but for the bodily help which has been rendered by a few domesticated animals.

From the point of view of the student of domesticated animals the races of men may well be divided into those which have and those which have not the use of the horse. [58] Although there are half a score of other animals which have done much for man, which have indeed stamped themselves upon his history, no other creature has been so inseparably associated with the great triumphs of our kind, whether won on the battle-field or in the arts of peace. So far as material comfort, or even wealth, is concerned, we of the northern realms and present age could, perhaps, better spare the horse from our present life than either sheep or horned cattle; but without this creature it is certain that our civilization would never have developed in anything like its present form. Lacking the help which the horse gives, it is almost certain that, even now, it could not be maintained.

We know the ancient natural history of the horse more completely than that of any other of our domesticated animals. We can trace the steps by which its singularly strong limbs and feet, on which rests its value to man, were formed in the great laboratory of geologic time. The story is so closely related to the interests of man that it will be well briefly to set it before the reader. In the first stages of the Tertiary period, in the age when we begin to trace the evolution of the suck-giving animals above the lowly grade in which the kangaroos and opossums belong, we find the ancestors of our mammalian series all characterized by rather weakly organized limbs fitted, as were those of their remoter kindred the marsupials, for tree climbing rather than for moving over the surface of the ground. The fact is, that all the creatures of this great clan acquired their properties of body in arboreal life, and with such relatively small and light bodies as were fitted for tree climbing. For this use the feet need to be loose-jointed, and so the system of five toes, each terminating in a sharp and strong nail or claw, [59] became fixed in the inheritances. When, gaining strength and coming to possess a more important place in the world, these ancient tree-dwellers were able to occupy the ground which of old had been possessed by the great reptiles, the limbs that had served well for an arboreal life had to undergo many changes in order to fit them for progression in the new realm.

If we watch the progress of a bear over the surface of the ground, we readily perceive how lumbering is its gait and how poor the speed which it attains. Its slow and shambling movement is due to the fact that it has the tree-climbing foot, and is not well fitted for motion such as is required in running. To attain anything like speed in this exercise it is necessary to support the body on the tips of the toes. Every man who has gained any skill in this art knows full well how incompetent he is if he tries to run with rapidity in the flat-footed manner. The bear cannot essay this method of progression on the toe-tips because its loose-jointed feet cannot be made to support its heavy body. In this way arose the necessity of developing a peculiar kind of foot when that part had to serve for rapid locomotion. The experiments to this end have been numerous and varied.

Thus in the elephants, which retain the originally numerous toes, the bones of these members are planted in an upright position and tied together with such strong muscles and sinews, that the foot parts have something like the solidity and strength of the upper portions of the legs. In the single-hoofed or horse-like forms, and in the cloven-footed animals, other series of experiments have been tried which in the end have proved most successful, giving us animals with the speediest movements of any animals except the creatures of the air.

A Hunter

[60]

The success which has been attained in our ordinary large herbivora, and which has made them competent to evade the chase of the beasts of prey, has been accomplished by reducing the number of the toes, giving the strength of the aborted parts to increase the power of those remaining. The result is the formation of two great groups, the double-hoofed forms, including the pigs, deer, cattle, sheep, and their kindred, and the single-toed species, of which our horse is the foremost example. In the reduction of the number of toes, different plans were followed in each of these groups. In the cloven-hoofed forms, a single toe [61] first disappeared, leaving but four; then the two outer of these were aborted, leaving two nearly equal digits. In the series of the horse, where we can trace the change more clearly, we find the earliest form five-toed, but the outer and inner digit shrunken so as to become of little use. This condition of the creature in the early Tertiaries gives us the beginning of the equine series, and shows that far away as the creature is now from ourselves, it originated from the main stem of mammalian life, from which our own forms have sprung. In the next higher stage in time, and likewise in development, we find these lessened toes at their vanishing point, and two of the remaining digits, lying on either side of what corresponds to the middle finger in our own hands, beginning to shrink in length and volume, while the central toe becomes larger and stronger than before. Last in the series we come to our ordinary equine form, in which nothing is left but the single massive extremity, though the remnants of two of the toes can be traced in the form of slender bones known as splints, which are altogether enclosed within the skin which wraps the region about the fetlock joints.

As if it were to show to us the history of this marvellous organic achievement, nature now and then, though seldom — perhaps not oftener than one in ten million instances — sends forth a horse with three hoofs to each leg. Two of these are small and lie on either side of the functioning extremity. Each of these hoofs is connected with a splint-bone which has in some way suddenly become reminded of its ancient use, and develops in a manner to imitate the creatures which passed from the earth millions of years ago. In most cases the splint-bones have no function whatever to perform. They [62] are indeed superfluous and injurious parts, and are likely from time to time to be worse than useless, becoming the seats of disease. In this beautiful instance, perhaps the fairest of all those showing how the highly developed forms of our time retain a memory of their ancestral life, we see how the advance in the series of the horse has been effected against the resistance ancient organic habit opposes to all gains. We can therefore the better understand how the building of the hoof represents the labor of geologic ages during which the slow-made gains were won.

In its present elaborate form, the hoof of a horse is the most perfect instrument of support which has been devised in the animal kingdom to uphold a large and swiftly moving animal in its passage over the ground. The original toe-nail, and the neighboring soft parts connected with it, have been modified into a structure which in an extraordinary manner combines solidity with elasticity, so that it may strike violent blows upon the hard surface of the earth without harm. The bones of the toe to which it is affixed have enlarged with the progressive loss of their neighbors of the extremity, until they fairly continue the dimensions of the bony parts of the leg. Moreover, they have lengthened out, so as to give the limb a great extension, and this, in turn, magnifies the stride which the creature can take in running. The result is that the horse can carry a greater weight at a swifter speed than any other animal approaching it in size.

[63-4]

On Rotten Row, Hyde Park, London

The needs which led, in a slow accumulative way, to the invention of the admirable contrivance of the horse's foot, were doubtless founded on the necessities of swift movement in fleeing from the great predaceous animals. Incidentally, however, as this development has gone on, the peculiarities [65] of the extremity have proved highly advantageous in defence, and the creatures have acquired certain peculiar ways of using their feet effectively to this end. The solid character of the hoof, its considerable weight, and the great power of the muscles of the hams, which are the principal agents in propelling the animal, make the hind feet capable of delivering a very powerful blow. The measure of its efficiency may be judged from the fact that a lion has been slain by a stroke from the foot of a donkey, and in their wild state a herd of horses with their heads together, can beat off the attack of the most powerful beasts of prey. In using the hind feet for assault or defence, horses have adopted an effective method of kicking which is unknown among other animals. Resting on their fore-legs, the hinder feet are thrown backward and upward, so that they may strike a blow six feet from the ground. Many of our cloven-footed animals have learned to strike cutting blows with the sharp hoofs of their fore-limbs—our bulls will stamp a fallen enemy with great force; but the backward kick of the horse is a peculiar movement, and is distinctly related to the peculiar structure of the animal's extremities.

It is an interesting fact that the development of a long and slowly elaborated series leading to the making of the horse appears to have taken place mainly, if not altogether, in the region about the headwaters of the Missouri River. In the olden days when this great work was done, that part of our continent was a well-watered country, much of its surface being occupied by great lakes which have long since disappeared. In the deposits accumulated in these bodies of fresh water are found the bones of the olden species telling the history of their series. It is not yet certain that the final [66] step of the accomplishment which gave us our existing species was effected in this land. It seems indeed most likely that the ancestral form of our domesticated horses found their way to the continents of the Old World, and there underwent the last slight changes, before they were made captive by man. If there ever were perfect horses on this continent, they had passed away from its area before the coming of man to the land. The history of our aborigines would have been quite other than it has been, if they had had a chance to win the assistance of this noble helpmeet.

Central Asia appears to have been the domicile of the horse when he first began his acquaintance with our kind. We do not know the original form of the creature. The wild horses existing at the present day in that part of the world, and which plentifully occur in other regions whereunto they have been taken by man, appear to have been set free from captivity.

Horse of a Bulgarian Marauder

The first domestication of the horse appears to have been brought about, at an early time in the history of our race, in northern Asia. The time when this feat was accomplished antedates our records. The creature may first have come into possession of the Tartar tribes, but it quickly passed over Asia and Europe and shortly became the mainstay of the Aryan and Semitic folk. None other of our domesticated forms has been disseminated with like rapidity, or at the outset with as little change in its original features. From the first the horse seems to have been mainly used as a saddle and pack animal. It has never served in any considerable measure for food. The failure to make use of the flesh of this animal appears to be common to most of the savage or barbaric people who keep horses, and has been transmitted [67] in a singularly definite way to all civilized folk. The origin of such a prejudice, despite the fact that the flesh of the horse is of excellent quality, can only be explained through the sympathetic motives common to all men. Their association with the horse, as with the dog, is so intimate as to make the use of these animals in the form of food more or less repugnant. In a small though unimportant way, mares have been used for milk, and there seems no reason to doubt that, if they had been carefully bred for this purpose, they might have been as serviceable as the cow. It may be that the failure to use the milk of the horse is to be accounted for on the same ground as the dislike to its flesh.

The horse was probably at first most valued for its use in war. The peoples which possessed it certainly had a great advantage over their less well provided neighbors. In fact the development of the military art, as distinguished from the mere fighting of savages, was made easy by the strength, endurance, fleetness, and measure of bravery characterizing this creature. In the wide range of species which have been domesticated or might be won to companionship with man, there is none other which so completely supplements the [68] imperfect human body, making it fit for great deeds. If the horse had been much smaller or larger than he is, he would have been far less serviceable to man. It was a most fortunate accident that the creature came to us with the proportions which insured a high measure of utility in various lines of activity. The elephant has been found too large for agricultural uses, and too powerful to be controlled by the will and force of his master under conditions of excitement.

Mare and Foal

Those peoples which early acquired the resources in the way of strength and fleetness which the horse put at their disposition, became inevitably the conquerors of the folk who were denied these advantages. If we consider the conditions which have led to the domination of the world by the Aryan and Semitic people, and the races which they have affiliated with them, we readily discern the fact that they have, to a [69] great extent, won by horse-power rather than by their own physical strength. Thus equipped by their able servants, they have pressed outward from their ancient realms and have in a way overridden the tribes which were unmounted.

So imposing is the effect of the horsed man on all peoples who are without previous knowledge of the united creatures, that it always carries fear to their hearts. To such folk the combination appears as a single terrible being. The ease with which the Spaniards conquered Mexico and Peru can, to a great extent, be attributed to the awe carried into the ranks of the savage footmen by their mail-clad horses. The Greeks, who were wont to represent the forces of nature and the accomplishments of man by skilfully constructed myths, have left a record showing their appreciation of the strength derived from the union of horse and man, in their fable of the Centaur, which possibly grew up in a time before their people had won the use of the animal, and when they only knew the creature by chance encounters with enemies who were mounted upon them. Although the naturalist of to-day perceives the impossibility of there ever having been on this earth a form uniting the trunk and fore-limbs of a quadruped to the upper part of a man's body, such scientific conceptions are a part of our modern, recently acquired store of knowledge. To the Greeks of the myth-making age the creature, half man, half horse, added but one more wonder to the vast store the world already contained. The currency of this fable shows us very clearly how great was the impression which the horse made upon primitive peoples.

To perceive the value of the horse in those ancient contests which opened the paths of civilization, we must note the fact that, until the invention of gunpowder, success in [70] breaking the ranks of an enemy depended mainly on the charge. With a large body of vigorous horsemen it was generally possible to overwhelm an enemy's line of battle, either by direct assault or by an attack on its flank or rear. If the reader is curious to see the value of horsemen in ancient warfare, he should read the story of the campaigns of Hannibal against the Romans in Italy. The first successes of that great commander—victories which came near changing the history of the western world—were almost altogether due to the strength lying in his admirable Numidian cavalry. The Romans were already good soldiers, their footmen more trustworthy than those which the Carthagenian general could set against them; but with his horsemen, as at Cannæ, he could wrap in the Roman line and reduce the most valiant legions to the confused herd which awaited the butcher.

[71-2]

Cavalry Horse

Although the invention of firearms has somewhat changed the conditions under which cavalry may be used, making indeed the direct charge more costly to the assailant than the assailed, it has in no wise diminished, but rather increased, the value of horses in military campaigns. In the line of battle horses have become necessary for the conveyance of field officers and messengers, and the right arm of battle, the artillery, could not possibly be managed except by horse-power. The swift marches of modern armies, by hastening the issue of contests, have spared the world half the woes of its great campaigns, and are made possible by the ready movement of supply trains, which could not be effected except by the help of these creatures. The result is that a large part of the military strength of any state rests not only in the valor and training of its fighting men, but in the supply of horses that its fields may afford. In this connection [73] it is instructive to compare the military strength of a country like China, where the horse is not a common element in the life of the people, with that of any of the western folk who may hereafter have to wrestle with that populous empire. Some writers, in their efforts to forecast the large politics of the future, have imagined that when the hardy and obedient Chinaman came to receive the European training in the military art, the armies of that country might prove from their numbers a menace to our own civilization. Such an issue seems in a high degree improbable, for the reason that the eastern realm could not provide the horses which would be necessary for the use of invading armies; nor is it at all likely that the rigid framework of their society will ever be so altered as to provide an abundance of these animals.

Plough Horses, France

Although in the first instance the horse served mainly, if [74] not altogether, as an ally of man in his contests with his neighbors, its most substantial use has been in the peaceful arts. As pack animal and drawer of the plough, the ox appears in general to have come into use before its swifter companion. The displacement of horned cattle has been due to the fact that their structure and habits make them much less fit for arduous and long-continued labor than the horse has been found to be. The cloven foot, because of its division, is weak. It cannot sustain a heavy burden. Even with the unincumbered weight of the body of the animal, the feet are apt to become sore in marches which the heavily mounted horse endures unharmed. Centuries of experience have shown that while the ox is an excellent animal for drawing a plough in a stubborn soil, and is well adapted to pulling carriages where the burden is heavy and the speed is not a matter of importance and the distance not great, the creature is too slow for the greater part of the work which the farmer needs to do. The pace which they can be made to take in walking is not more than half as great as that of a quick-footed horse moving in the same gait; and the ox is practically incapable, because of its weak feet, of keeping up a trot on any ordinary road.

But for the fact that an aged ox may be used for beef, they would doubtless long since have ceased to serve us as draught animals. As it is, with the growing money value of the laborer's time, this slow-moving creature is steadily and rather rapidly disappearing from our farms. This change, indeed, is one of the most indicative of all those now occurring in our agriculture. It is an excellent example of the operations which the increase in the workman's pay is bringing into our civilization.

The natural advantages of the horse for the use of man [75] consisted in its size, strength, and endurance to burden; form of the body, which enabled a skilful rider to maintain his position astride the trunk; and the peculiar shape of the mouth and disposition of the teeth which made it possible to use the bit. With these direct physical advantages there were others of a physiological and psychic sort, of equal value. The creature breeds as well under domestication as in the wilderness; the young are fit for some service in the third year of their life, and are, at least in the less elaborated breeds, in a mature condition when they are five years old. Experience shows that the animal can subsist on a great variety of diet, being in this regard surpassed only by its humbler kinsman the donkey, and by the goats. There are few fields so lean that they will not maintain serviceable horses. They do well alike in mountain pastures and amid the herbage of the moistest plainland.

The mental peculiarities of the horse are much less characteristic than its physical. It is indeed the common opinion, among those who do not know the animal well, that it is endowed with much sagacity, but no experienced and careful observer is likely to maintain this opinion. All such students find the intelligence of the horse to be very limited. It requires but little observation to show that the creature observes quickly, and in some way classifies the objects with which it comes in contact. The fear aroused in it by unknown things makes this feature of attention to the surrounding world very evident. Almost all these animals retain a tolerably distinct memory of the roads which they have traversed, even if they have passed over them but a few times. The studies which I have made on this point show me that the average horse will be able to return on a road [76] which it has traversed a few hours before, with less risk of blundering than an ordinary driver. Some well-endowed animals can remember as many as a dozen turnings in a path over which they have journeyed three or four times. It seems almost certain that their guidance in these movements is not at all effected by the sense of smell, but is due to a distinct memory of the detailed features of the country.

Belgian Fisherman's Horse

Good as is the horse's memory, it is difficult to organize its actions on that basis. Only in rare cases and with much labor can he be taught to execute movements that are at all complicated. Fire-engine horses may be trained of their own will to step into the position where they are to be attached to the carriage. Some artillery horses will, as I have noticed, associate the sound of the bugle with the resulting movements of the guns and take the appropriate positions, where they may be out of danger in the rapid swinging of the [77] teams and carriages. It is partly because of this training received by disciplined artillery horses, that it seems to many experienced officers not worth while to have militia companies in this arm, who have to manœuvre with animals untrained for the service. Although some part of this mental defect in the horse, causing its actions to be widely contrasted with those of the dog, may be due to a lack of deliberate training and to breeding with reference to intellectual accomplishment, we see by comparing the creature with the elephant, which practically has never been bred in captivity, that the equine mind is, from the point of view of rationality, very feeble.

The emotional side of the horse's nature seems little more developed than its rational. Although they have a certain affection for the hand which feeds them, and in a mild way are disposed to form friendships with other animals, they are not really affectionate, and never, so far as I have been able to find, show any distinct signs of grief at separation from their masters or of pleasure when they return to them. Although there are many stories appearing to indicate a certain faithfulness in horses which have remained beside their fallen and wounded riders, the facts do not justify us in supposing that such actions are due to the affection a dog clearly feels.

Horses for Towing on the Beach in Holland

We have been singularly led astray by a chance use of the epithet "horse," which has come to be applied to many organic forms and functions where strength is indicated. Thus, in the case of plants we speak of "horse-radish" or "horse-mint," denoting thereby spices which have strong qualities. Horse-chestnut is another instance of the application of the term to plants. It chanced that "horse-sense" [78] came to be used to indicate a sound understanding, and in an obscure way, but in a manner common with words, this has led to a vague implication of mental capacity in the animals whence the term is derived. The fact is that our horses, as far as their mental powers are concerned, appear to be the least improvable of our great domesticated animals.

[79-80]

A Hurdle Jumper

Little elastic as the horse appears to be on the psychic side of its nature, in its physical aspects it is one of the most plastic of all the forms subjected to the breeder's art. It requires no more than a glance at the streets of our large cities to see how great is the range in size, form, and carriage of these animals which may be found in any of our great centres of civilization. We readily perceive that these variations have a distinct relation to the several divisions of human activity in which this creature has a share. The massive [81] cart-horse, weighing it may be as much as eighteen hundred or two thousand pounds, heavy limbed, big headed, unwilling to move at a pace faster than a slow trot, yet not without the measure of beauty seemingly inseparable from the species, contrasts very markedly with the alert saddle animal bred for speed and grace, and for the easy movement which makes it comfortable to the equestrian. Between these extremes we may note minor differences which, though they may not strike those persons who take only a commonplace view of the creatures, are most marked to the initiated. The trotter, the coach horse, the strong but nimble animals which are used in fire-engines and other heavy carriages which have to be swiftly moved, mark the results of breeding designed to insure particular qualities, and show how readily the physical features of the animal can be made to fit to our desires.

Although from an early day a certain amount of care has been given to breeding horses for saddle purposes, the careful and continuous choice which has led to the modern variations is a matter of only a few centuries of endeavor. So far as we can judge from the classic monuments, the olden varieties were mere varieties of the pony—the small, compact, agile creature which had not departed far from the parent wild form. It seems to me doubtful whether any of the horses possessed by the Greeks or Romans attained a weight much exceeding a thousand pounds, or had the peculiarities of our modern breeds. The first considerable departure from the original type appears to have been brought about when it became necessary to provide a creature which could serve as a mount for the heavy armored knights of the Middle Ages, where man and horse were weighted with from one to two [82] hundred pounds of metal. To serve this need it was necessary to have a saddle animal of unusual strength, weighing about three-quarters of a ton, easily controllable and at once fairly speedy and nimble. To meet this necessity the Norman horse was gradually evolved, the form naturally taking shape in that part of Europe where the iron-clad warrior was most perfectly developed. In the tapestries and other illustrative work of that day, when the knight won tournaments and battle-fields, gaining victory by the weight and speed which he brought to bear upon his enemies, we can see this splendid animal, in physical form, at least, the finest product of man's care and skill in the development of the lower species.

With the advance in the use of firearms the value of the Norman horse in the art of war rapidly diminished. This breed, however, has, with slight modifications, survived, and is extensively used for draught purposes where strength at the sacrifice of speed is demanded. It is a curious fact that the creatures which now draw the beer wagons of London often afford the nearest living successors in form to the horses which bore the mediæval knights. It is an ignoble change, but we must be grateful for any accident which has preserved to us, though in a somewhat degraded form, this noblest product of the breeder's art, which, even as much as the valor of our ancestors, won success for our Teutonic folk in their great struggle with Islam. A tincture of this Norman blood, perhaps the firmest fixed in the species of any variety, pervades many other strains most valuable in our arts. The best of our artillery horses, particularly those set next the wheels, are generally in part Norman. In the well-known American Morgan, the swiftest and strongest of our [83] harnessed forms, the observant eye detects indications of this masterful blood.

The Norman strains of horses retain certain interesting indications of their ancient lineage and occupation. As appears to be common with old breeds, the stock is readily maintained. It breeds true to its ancestry, with little tendency to those aberrations so common in the newly instituted varieties. When crossed with other strains, the effect of the intermixture of this strong blood is distinctly traceable for many generations. In their mental habits these creatures still appear to show something of the effects of their old use in war; it is a valiant race, less given to insane fear than other strains, and, even under excitement, more controllable than the most of their kindred. So far as I have been able to learn, they seem singularly free from those wild panics which are so common among our ordinary horses. It does not seem to me fanciful to suppose that these qualities were bred in the stock during the centuries of experience with the confusion of battle-fields and tournaments.

Exercising the Thoroughbreds

The horse, in common with the other domesticated animals varying readily in the hands of the breeder, undergoes a certain spontaneous change which in a way corresponds to the physiography of the region in which it is bred. At first sight it may seem as if these alterations are due to the admixture of previously existing varieties, or to the institution of peculiarities by some process of selection. I am, however, well convinced that these variations are in good part due to a direct influence from the environment. Thus in our high northern lands there is a distinct and spontaneous reduction in size of the creatures, which attains its farthest point in the Shetland pony. Again, as we go toward the [84] tropics, a like though less conspicuous decrease in bulk is observable. The largest animals of the species develop in the middle latitudes, the realm where the form appears to have acquired its characters. The speed with which these local variations are made is often great. Thus the horses of Kentucky have, in about a century, acquired a certain stamp of the soil which makes it possible, in most cases, for the observer to identify an individual as from that State, though he may find it in a field a thousand miles away. The defining indications are not limited altogether to bodily form, but are shown in what might seem trifling features of carriage and behavior. The difference between the horses of Great Britain and those of the United States seems to me, from repeated observations, to be quite as great as that separating the men of the two realms. I believe that if a lot of a thousand, taken in equal parts from either land, were put together, a person well accustomed to taking account of [85] these animals could separate them into two herds, with less than ten per cent. of error. It is doubtful if a more perfect selection could be made if the same experiment were tried on an equal number of men, provided the indices to be derived from peculiarities of speech or dress could be excluded.

An Arabian Horse

By some the Arabian horse is thought to be the most remarkable specialization of the kind which has been attained. In his native country and in his perfection, the Arab breed has been seen by but few persons who have been specially trained in noting the peculiarities of the animal. So far as I have been able to judge by pictures and a few specimens, said to be thoroughbreds of their stock, which I have had a chance to see, the Arabian form of the horse appears to have been led less far away from the primitive stock than many of our European and American varieties.

Arabian Sports

[86]

The very great, if not the preëminent, success of the horse in Arabia is the more remarkable from the fact that it has been attained under conditions which, from an *a priori* point of view, must be deemed most unfavorable. This variety has been bred in a land of scant herbage and deficient water-supply, where the creature has had from time to time, indeed we may say generally, to endure something of the dearth of food which stunts the Indian ponies and the other horses of the Cordilleran district. The ancestors of the horse appear to have attained their development in well-watered and fertile regions. All the varieties bred within the limits of civilization do best on rich pasturages such as Arabia does not afford. The success of the horse in that land shows how devoted must have been the care which has been given to its nurture. Fitting, as the Arabian horse does, exactly to the needs of nomadic people engaged in almost constant warfare, it has [87] naturally been a far more important helper to the wild folk of the desert lands about the eastern Mediterranean and the Red Sea than to any other race. In those lands horses fell into the keeping of a very able folk. The contrast between the care devoted to the animals by them, and that which our Indians give to their ponies, is a fair measure of the difference in the ability of these very diverse races.

As a whole, the horse demands for his best nurture and keeping an amount of care required by no other animal which has been won to the uses of man, unless perhaps it be the silkworm. Kept in its best state, the horse has to be sedulously groomed. To be maintained in its very best condition some hours of human labor must each day be given to keeping his skin in order. The effect arising from a friction on the horse's hide is not confined to the beauty that comes from cleanliness, but in a curious way reacts upon the general nervous tone of the animal. All those who are familiar with horses will, I think, agree with me that much grooming distinctly increases the endurance and elasticity of their bodies. The influence of the grooming process appears to be somewhat like that obtained by massage and friction of the skin in the training of an athlete. More than once I have had occasion to observe the effect of this process on some ancient horse of good blood, which for years had been allowed in its old age to go uncared for as an idle tenant of the pastures. Two or three days of assiduous grooming will bring back the strength and suppleness to the aged limbs, and restore something of the olden spirit. The effect obtained from this care is the more remarkable for the reason that nothing similar to it was experienced by the wild ancestors of these creatures. It is as artificial as bathing in the case of man. The influ [88] ence of the treatment shows how very unnatural is the state of our civilized horses.

The task of providing horses with food is more considerable than in the case of any of our other domesticated creatures. By nature the animal is a frequent feeder, and does not well endure long fasts. Its stomach is rather small for the size of the body, and the digestive process appears to be more than usually rapid. A mounted animal, when taxed to its utmost, should be fed four or five times a day, and with less than three good meals is apt to break down. No such care in the matter of provender is necessary in the case of the other members of man's animal family. The contrast between the physiological conditions of the camel and those of the horse are fully recognized by the Arabs, in their almost complete neglect of the individuals of the one species and their exceeding care of the other.

[89-90]

English Polo Ponies

Perhaps the greatest element of care which man has had to devote to the horse is found in the matter of shoeing. In the state of nature the admirably constructed hoof sufficiently provided the animal against the excessive wearing of its horny extremity. Nature, however, rarely provides for more strength and endurance than the creature in its wild state demands; and so it comes about that when horses have to bear burdens or draw carriages, particularly on roadways, their unprotected feet will not withstand the strain which is put upon them, the rate of growth of the structure composing the hoof not being sufficiently rapid to make good the wearing which these unnatural conditions impose. For thousands of years, in the roadless stages of man's development, the difficulties arising from the wearing of the hoof were not serious, for the creatures trod either on turf-covered [91] plains or on the soft ways of the desert. When the advance of culture made roads necessary, when carriages were invented and something like our modern conditions were instituted, it became imperatively necessary to provide additional protection for the feet. We find the Greeks, in the classic time, wrestling with this problem. Xenophon, in his treatise on the care of horses, advises that they be reared on stony ground, he having observed that, in a natural way, the hoof becomes somewhat adapted to the necessities of its conditions. The Romans found the difficulty from the tender foot of the horse yet more serious on their paved roads; but both these classic people showed, in their ways of dealing with the difficulty, that lack of inventive skill which so curiously separates the olden from the modern men. They devised soles of leather and bags as coverings for the horse's feet, but none of the contrivances could have been very serviceable. All such coverings must have been quickly worn out in active use.

So far as we can determine, it was not until about the fourth century of our era that the iron horseshoe was invented. This valuable contrivance appears to have originated in Greek or Roman lands, probably in the former realm, for it first bore the name of "selene," from its likeness to the crescent shape of the new moon. Although simple, the horseshoe was a most important invention, for it completely reconciled the animal to the conditions of our higher civilization by removing the one hinderance to its general use in the work of war and commerce. It is probable that with this invention began the great task of differentiating the several breeds of European horses for their use in various employments, as draught animals for packing purposes, as [92] light saddle horses, and the bearing of armored men. Neither the draught nor the war horses of Europe could well have been specialized until their heavy bodies were separated from the ground by these metallic coverings of the hoof.

Syrian Horse

Much has depended on the specialization of the horse into different breeds, made possible by the iron shoe. By reconciling the creature to uses—agriculture, which depends on draught animals, and the commerce of importance, which can only be effected by means of wagons—the rapid economic development of our civilization was made possible. By developing a horse capable of bearing an armored man, Europe was brought into a condition in which organized armies took the place of mere forays, and so the development of centralized states was promoted. In the warfare between the Mohammedans and the Christian states of Europe, in [93] the campaigns with the Turks and the Saracens, it is easy to see that the powerful breeds of horses reared in western and northern Europe were a mighty element in determining the issue of the contest. The battles of these momentous campaigns represented, not only a struggle between the Christian Aryans and the Semitic followers of Mahomet, but, in quite as great a degree, the war was waged between the light and agile steeds of the Orient and the massive and powerful animals that bore the mail-clad warriors of the West. On the field of Tours, when the fate of Christian Europe for hours hung in the balance, we may well believe that the strong and enduring horses of the northern cavalry did much to give victory to our race.

Along with our general account of the place of the horse in civilization, it is fit to give something to the story of his near, though inferior, kinsmen, the ass and the mule, both of which have played a subordinate, though important, part in the same field of endeavor in which the nobler species has done so much for man. The original progenitors of our donkeys differed from the ancestral form of the horse by variations of good specific value. So far as we can determine from visible features, these forms were more distinctly parted than the dog and the wolf, or either of these animals from the jackal. Nevertheless, these equine forms are clearly closely akin, for they may be bred together. Although the original stock of the ass may possibly have been lost, it seems most likely that the wild forms which exist in Asia have not wandered off from captivity, but are the remnants of the original wilderness form.

It appears likely that the two domesticated equine species have been under the care of man for about the same length [94] of time; but the difference in their condition, and in the place which they hold in civilization, is very great. As we have seen, the horse has been made to vary in a singular measure, its form and other qualities changing to meet the need or fancy of its master. Its humbler kinsman has remained almost unchanged. Except small differences in size, the donkeys in different parts of the world are singularly alike. In part this lack of change may be explained by the relative neglect with which this species has been treated. From the point of view of the breeder it has perhaps been the least cared for of any of our completely domesticated animals. In some parts of the world, as for instance in Spain, where a long-continued effort has been made to develop the animal for interbreeding with the horse, the result shows that the form is relatively inelastic. It is doubtful if any conceivable amount of care would develop such variations as the horse now exhibits.

The principal hinderances to the general acceptance of the donkey as a help-meet to man are found in its small size and slow motion. These qualities make the creature unserviceable in active war or in agriculture, and they seem to be so fixed in the blood that they are not to any extent corrigible. So long as pack animals were in general use, and in those parts of the world where the conditions of culture cause this method of transportation to be retained, the qualities of the donkey have proved and are still found of value. The animal can carry a relatively heavy burden, being in such tasks, for its weight, more efficient than the horse. It is less liable to stampedes. It learns a round of duty much more effectively than that creature, and can subsist by browsing on coarse herbage, where a horse [95] would be so far weakened as to become useless. Thus, in developing the mines in the unimproved wilderness of the Cordilleras, where ores of the precious metals have to be carried for considerable distances, trains of "burros" are often employed. The animals quickly learn the nature of their task, and will do their work with but little guidance from man.

In general we may say that the donkeys belong to a vanishing state of human culture, to the time before carriage-ways existed. Now that civilization goes on wheels, they seem likely to have an ever-decreasing value. A century ago they were almost everywhere in common use. At the present time there are probably millions of people in the United States to whom the animal is known only by description. In a word, the creature marks a stage in the development of our industries which is passing away as rapidly as that in which the spinning-wheel and the hand-loom played a part.

As the use of the ass in the economic arts began to decline, the mule or hybrid progeny of this creature and the horse has progressively increased. Although the value of this mongrel has been known, particularly in southern Europe, from very early days, its most extensive employment has been found in the old slave-holding States of the Federal union. The custom of using mules has been almost unknown in England, and has never been generally adopted in the northern part of the United States. It appears to have been introduced into southern regions by the Spaniards and the French, and there to have spread, because of the peculiar fitness of the creature to the climate and the employment it had to endure in that part of America. The mule has the [96] peculiar advantage that it is on the average as large as the horse, is nearly as quick-footed when walking, and has at the same time a considerable share of the patient endurance to hard labor and scant fare which characterizes the donkeys. It matures somewhat more speedily than its nobler kinsman, being ready to meet severe strains perhaps a year earlier. Unless unconscionably abused, its period of fitness for hard work endures about one-third longer, often lasting for thirty years. It is singularly exempt from disease, its sturdy frame withstanding rude usage until the old age time.

In the Circus

The mule is especially interesting to the naturalist for the reason that it affords the only certain case in which a hybrid has proved decidedly serviceable to man. It is not unlikely that a similar mixture of the blood of two species occurs in our ordinary cats, and it may exist in the case of the dog [97] and in some of the domestic birds; but so far as we know, there has been no other useful result from the hybridizing, if it has occurred. Moreover, the mule is unique for the fact that the animal is distinctly stronger for its weight, and more enduring than either species which his blood combines. In fact, there is no product of man's industry in relation to domesticated animals which is more interesting than this singular creature. At present, its use appears to be going out of vogue; the evidence goes to show that the hybrid has no place in the affections of mankind, and that it is only likely to be kept in its use in tropical countries, and particularly in regions where the beasts have to be under the care of slaves or other negligent folk. It is a singular fact in connection with this hybrid, that it is nearly absolutely sterile, there being only two or three cases on record in which they have proved fecund. It seems, however, possible that if these rare instances of continued breeding were to be duly used, an intermediate species might be permanently established. This is, indeed, one of the most important lines for experiment which could be undertaken by an institution devoted to the study of problems relating to domestication.

It is commonly thought that a mule is a stupider creature than the horse; but I have never found a person, who was well acquainted with both animals, who hesitated to place the mongrel in the intellectual grade above the pure-blood animal. There is, it is true, a decided difference in the mental qualities of the two creatures. The mule is relatively undemonstrative, its emotions being sufficiently expressed by an occasional bray — a mode of utterance which he has inherited from the humbler side of his house in a singularly unchanged way. Even in the best humor it appears sullen, and lacks [98] those playful capers which give such expression to the well-bred horse, particularly in its youthful state. It is evident, however, that it discriminates men and things more clearly than does the horse. In going over difficult ground it studies its surface, and picks its way so as to secure a footing in an almost infallible manner. Even when loaded with a pack, it will consider the incumbrance and not so often try to pass where the burden will become entangled with fixed objects.

Mules soon learn the difference between those who have the care of them and strangers. It is a well-known fact that trouble awaits the wight who unwarily ventures to take from the stall a mule which has not the advantage of his acquaintance. On this account they are rarely stolen. Even in the daytime they are often dangerous for strangers to approach, and the most of the ill-usage which men receive from their heels arises where unwitting people venture to treat them as they would horses. Mules are much less liable to panic-fear than the most of our domesticated animals, yet, when kept in the herded way, they occasionally become stampeded. Many a soldier of our Civil War, where mules played a large part in the campaigns, doubtless remembers the mad outbreaks of these creatures from their corrals, when they went charging through the army with a fury which, if directed against an enemy, would have been almost as effective as a cavalry charge.

It is interesting to note that mules have a greater disposition to adopt a leader in their movements than we note in either of the species whence they come. In the old days when mules were plentifully bred in Kentucky, and taken thence for sale to the plantation States, they went forth in droves, commonly under the leadership of a bell horse, or, by [99] preference, a mare, which it was quite the custom to choose of a white color. In the course of a few hours the creatures would learn to know their guide, and to follow the leader with so little trouble that two men could conduct a throng of several hundred. Nevertheless, if the foremost mule of the procession turned aside, all the others would blindly follow him in the manner of a flock of sheep.

I recall an amusing instance of this "follow-my-leader" motive which occurred many years ago in a way somewhat personal to myself, in southern Kentucky. Engaged in survey work, I was passing along a quiet road when in the distance I heard a thunder of hoofs, and in a moment saw a great drove of mules, the appointed leader of which, a man on a white horse, had fallen to the rear of the column. The creatures, thinking that it was their duty to overtake the missing master, were going on the full run. Heeding the shouts of the troubled herder, I turned my wagon across the road, which, being at that point very narrow, was effectually barricaded by the vehicle. Although the rush was so wild that the brutes nearly overset my "outfit," they were brought to a full stop. Unhappily, on one side of the road and one hundred feet or so from it, there was a comfortably built southern house, with a broad gallery extending along the front; while in the door of the mansion were some women who had been attracted by the tumult. No sooner had the mob of mules been brought to a state of surging quiet, than one of the creatures jumped the picket fence, and started for the open house-door, thinking, perhaps, that he would find some peace of life in what probably seemed to him his accustomed barn. In much less time than it takes to tell it, a hundred or more mules were on the gallery, the floor of which gave way beneath [100] their weight; they quickly broke down the columns which supported the roof, so that the whole structure at once became a heap of wood and mules. The unhappy proprietor of the drove, in his consternation, forgot even to swear—an art which I have never known on any other occasion to pass from a mule-driver; and, sitting on his white horse, he lifted his hands like an oriental in prayer, and said to me meekly, "Did you ever in all your life?" I assured him that I had never, and went my way, leaving him to settle an interesting case of damages with the owner of the mansion.

In considering the general influence of the horse and its kindred forms on human culture, we clearly perceive that we are now attaining a time when the machinery of civilization is to depend in a much less degree than of old on the help which these creatures give to man. Even fifty years ago the horse was far more necessary to the work of our kind than it is at present. Going back a hundred years, we perceive that the population of the civilized world could not possibly have been maintained, if by some disease all the horses had been swept away. Such a calamity in the year 1800 would have led to the depopulation of almost all the cities of the interior country, famine would have ravaged our States, and the whole economic system of society would have had to be reconstructed. Now the greater part of the work which of old had to be done by horses, can, at a slight increase of cost, be effected by mechanical engines. Ploughing, except on steep hillsides and in very stony ground, can be cheaply and effectively done by steam. The same agent can propel the harvesters and work the threshing machines. Even farmers who till fields of no great extent find it desirable to do much of their work by steam-engines, for [101] the reason that fuel is less costly than horse feed. An interesting instance to show how far mechanical inventions have taken the place of horsed wagons in the work of civilized communities was afforded by the horse distemper which swept over the country in 1872. During the week or more in which this epidemic was at the worst, the State of Massachusetts was practically unhorsed, yet the greater part of the necessary business, that required to bring provisions to the town, was effected by means of the railways. The same incident shows, however, in another way, how absolutely necessary this animal is, in certain parts of our work. For the great Boston fire, which occurred at that time, was doubtless due to the fact that, owing to the sickness of the horses, an effort was made to drag the engines by hand-power, with the result that they came upon the ground so slowly as to give the fire a chance to become an uncontrollable conflagration.

In the present state of our arts there is one great occupation which we cannot conceive to be carried on without the services of horses. This is war. It is hardly too much to say that all our highly elaborated military system has depended for its development, as it does for its maintenance, on the transportation value of horses. Much has been said of late as to the use of bicycles as adjuncts to armies, and in a certain limited way they will doubtless prove serviceable in future campaigns; but no one who has had any experience of military duty, with its work across tilled fields and through forests, can imagine a man on a wheel rendering any very effective service except under peculiar conditions. Moreover, no ordnance corps can do its appointed work in the rear of a line of battle without sending its wagons across [102] country and over ground which no unhorsed vehicle could traverse.

The mark of the old utility of the animal in varied employment is retained in our use of the term horse-power in measuring the energy of engines. That gauge of strength of old determined what man could do in the severest taxes upon the forces at his command. In attaining the point where, owing to the possession of horses, he could use this standard, he won a great way beyond the station of his ancestors, who had but the strength of men at their command. Modern invention, by giving us heat-engines, has made the way for an advance. In another century, or even in another generation, the horse may, save for the uses of war, be confined to the position of a luxury and an ornament.

[103]

THE FLOCKS AND HERDS: BEASTS FOR BURDEN, FOOD, AND RAIMENT

Effect of this Group of Animals on Man.—First Subjugations.—Basis of Domesticability.—Horned Cattle.—Wool-bearing Animals.—Sheep and Goats.—Camels: their Limitation.—Elephants: Ancient History; Distribution; Intelligence; Use in the Arts; Need of True Domestication.—Pigs: their Peculiar Economic Value; Modern Varieties; Mental Qualities.—Relation of the Development of Domesticable Animals to the Time of Man's Appearance on the Earth.

It is not too much to say that the opportunity to go forward on the paths of culture, at least the chance to advance any considerable distance beyond the estate of primitive men, depends in a considerable measure upon what the wilderness may offer in the way of domesticable beasts of burden. Where such exist we find that the folk who dwell with them in any land are almost certain to have made great advances. Where the surrounding nature, however rich, denies this boon, we find that men, however great their natural abilities may appear to be, exhibit a retarded development. Thus in North America, where there was no domesticable beast of burden, the Indians, though an able folk, remain savages. So, too, in central and southern Africa, where the mammalian life, though rich, affords no large forms which tolerate captivity, the people have failed to attain any considerable culture. On the other hand, in the great continent of the Old World, where the horse, the ass, the buffalo, the camel, and the elephant existed in the primitive wilds, men rose swiftly toward the civilized station.

Domesticated Buffaloes in Egypt

[104]

The immediate effect arising from the possession of beasts of burden is greatly to enlarge the scope and educative value of human labor. A primitive agriculture, sufficient to provide for the needs of a people, can be carried on by man's labor alone, though the resulting food-supply has generally to be supplemented by the chase. Rarely, if ever, are the products of the soil thus won sufficient in quantity to be made the basis of any commerce. Such conveyance as is necessary among the people who are served by their own hands alone, has to be accomplished by boat transportation or by the backs of men. The immediate effect of using beasts for burden is the introduction of some kind of plough, which spares the labor of men in delving the ground, and the use of pack animals, which, employed in the manner of caravans, greatly promotes the extension of trade. A great range of secondary influences is found in the development of the arts of war, by which people who have become provided with pack or saddle ani [105] mals are able to prevail over their savage neighbors, and thus to extend the realm of a nascent civilization. Yet another influence, arising from the domestication of large beasts, arises from the fact that these creatures are important storehouses of food; their flesh spares men the labor of the chase, and so promotes those regularities of employment which lead men into civilized ways of life. In fact, by making these creatures captive, men unintentionally brought themselves out of their ancient savagery. They were led into systematic and forethoughtful courses, and thus found a training which they could in no other way have secured.

Cattle of India

The first and simplest use made of the animals from which man derives strength appears to have been brought about by the subjugation of wild cattle—the bulls and buffaloes. Several wild varieties of the bovine tribe were originally widely disseminated in Europe and Asia, and these forms must have been frequent objects of chase by the ancient hunters. Although in their adult state these animals were doubtless originally intractable, the young were mild-mannered, and, as we can readily conceive, must often have been led captive to the abodes of the primitive people. As is common with all gregarious animals which have long acknowledged the authority [106] of their natural herdsmen, the dominant males of their tribe, these creatures lent themselves to domestication. Even the first generation of the captives reared by hand probably showed a disposition to remain with their masters; and in a few generations this native impulse might well have been so far developed that the domestic herd was established, affording perhaps at first only flesh and hides, and leading the people who made them captives to a nomadic life—that constant search for fresh fields and pastures new which characterizes people who are supported by their flocks and herds.

It is a curious fact that the kindred of the buffaloes and bisons differ exceedingly in the measure of their domesticability. Thus, the ordinary buffalo of Asia, though a dull brute, is very subjugable, even in the literal sense, for he makes a tolerable beast for the plough and bears the yoke with due patience. His African kinsman, on the other hand, is perhaps the most unconquerable of all the large wild animals. The late Sir Samuel Baker, in answer to my question as to what wild form was the most to be feared in combat, unhesitatingly answered, "The African buffalo, the bulls of which charge home upon any aggressor with an immediate and determined fury, which often enables them to kill the hunter after they have been shot through the brain." Our American bison, though a much milder-spirited beast, seems also to be essentially undomesticable for the reason that he cannot be taught to subordinate his desires to the will of man. He can readily be brought to the point where he will tolerate captivity; but if, when engaged in ploughing, it occurs to him that he needs water, he will straightway go in search of it, not in a vicious, but in a perfectly obdurate manner. This quality of mind appears to be accountable for the failure of the many [107] experiments which have been made to domesticate this interesting American form.

The limitations of the domesticating work, the fact that as between two kindred species the one has been chosen by man and the other left, indicate the truth—which is generally of much importance—that the intellectual qualities of animals commonly differ more than their frames. This is a part of the larger fact that with the advance in organization the individuality, as regards the whole spiritual field in persons and species alike, becomes greater. The culmination of the tendency is seen in man, where, with bodies which do not vary much, we have an almost infinite range in individual qualities.

This is perhaps a good place in which to make answer to the suggestion that the domesticability of the animal species is in inverse proportion to their native courage and independence of mind. The reader will see how fallacious is this common notion if he will consider the quality of the supremely domesticated creature, the dog. There is probably no beast which has a larger share of natural courage and of independent motive. When not under the control of their masters, they have perhaps as free a contact with nature as any creature in the world; the same thing may be said of the elephant, which, next to the dog, lends himself most obediently to the requirements of the master. Owing to the power of his huge body and to the ease with which he wins his food, he is in his native wilds the least dependent of land animals. Except from the assaults of man, he has nothing to fear; yet when enslaved he at once surrenders himself to his captors. In general, it may be said that the true gauge of domesticability is the sympathetic motive, that strange outgoing spirit which leads the mind to recognize the life about it and to accept [108] that life as a part of its own. In other words, the domesticability of man is due to his willingness to enter into social relations and rests on the same foundation that supports his intercourse with the lower animals he has won to his use.

Indian Bullock and Water-Carrier

It is probable that the first use which was made of beasts of burden, in ways in which their strength became useful to man, was in packing the tents and other valuables of their masters as they moved from place to place. Even to this day in certain parts of the world bulls and oxen serve for such purposes. In fact the nomadic life, a fashion of society which is enforced wherever people subsist from their cattle alone, leads inevitably to such use of the beasts. In the southern Appalachian district of this country there remain traces of this service rendered by bulls and oxen. These creatures, provided with a kind of pack saddle, are occasionally used in conveying the dried roots of the ginseng, beeswax, feathers, and the peltries which are gathered by the inhabitants of remote districts, not accessible to carriages, to the markets of the outer world. All the varieties of ordinary cattle could be made to serve as burden-carriers, and they doubtless would be continued to be [109] used for saddle purposes in one way or another but for the wide use of the horse, a creature very much better adapted for carrying weight. The cloven foot of the bulls and buffaloes gives a weakness to the extremities which will quickly lead to disease in case they are forced to carry heavy loads such as the horse or ass may safely bear.

Ploughing in Syria

The help which our bovine servants afford us by the power which they exert in traction, as in drawing ploughs, sleds, or wagons, appears to have been first rendered long after their introduction to the ways of man. The first of these uses in which the drawing strength of these animals was made serviceable appears to have been in the work of ploughing. In primitive days and with primitive tools, hand delving was a sore task. The inventive genius who first contrived to overturn the earth by means of the forked limb of a tree, shaped in the semblance of a plough and drawn by oxen, began a great revolution in the art of agriculture. To this unknown genius we may award a place among the benefactors of mankind, quite as distinguished as that which is occupied by the equally unknown inventors of the arts of making fire or of smelting ores. After the experience with the strength of oxen had been won from the work of ploughing, it was easy [110] to pass to the other grades of their employment, where they were made to draw carriages.

Next after the contribution which the kindred of the bulls, have made by their strength, we must set that which has come from their milk. Although this substance can be obtained in small quantities from several other domesticated animals, the species of the genus Bos alone have yielded it in sufficient quantities greatly to affect the development of man. It is difficult to measure the importance of the addition to the diet, both of savage and civilized peoples, which milk affords. It is a fact well known to physiologists that in its simple form this substance is a complete food, capable when taken alone of sustaining life and insuring a full development of the body. It is indeed a natural contrivance exactly adapted to afford those materials which are required for the development and restoration of creatures essentially akin to our own species. Those races which avail themselves extensively of it in their dietary are the strongest and most enduring the world has known. The Aryan folk are indeed characteristically drinkers of milk and users of its products, cheese and butter. It may well be that their power is in some measure due to this resource.

[111-2]

Winnowing Grain in Egypt

In our horned cattle man won to domestication creatures which were admirably suited to promote his advancement from savagery to civilization. Indeed, the possession of these animals appears to have been a prime condition of his advancement. With them, however, as with the camel, there came little in the way of those sympathetic qualities which have made it possible for our race to establish affectionate relations with other captive forms. Long intercourse with man has, it is true, somewhat diminished the wildness of these creatures, though the males remain the most indomitably fero [113] cious of all our servants. The truth seems to be that the bovine animals have but little intellectual capacity, and it has in no wise served the purposes of man to develop such powers of mind as they have. We have ever been given to asking little of them, save docility. This we have in a high measure won with our milch cows, which of all our domesticated creatures are perhaps the most absolutely submissive; the more highly developed of them being little more than passive producers of milk, almost without a trace of instincts or emotions except such as pertain to reproduction and to feeding. It is a noteworthy fact that in all the great literature of anecdote concerning our domesticated animals, there is hardly a trace of stories which tend to show the existence of sagacity in our common cattle.

It is evident that the variability of our domesticated bovines, as far as their bodies are concerned, is very great. Between the ancient aurochs and the more highly cultivated of its descendants, the difference is as great as that which separates any other of our captive animals from their wild ancestors. In size, shape, in flesh-and milk-giving qualities, the departure from the old form of the wilderness is remarkable. Moreover, at the present time these diverse breeds of horned cattle are rapidly being multiplied, the distinctive forms probably being twice as numerous as they were at the beginning of the present century. The process of selection has led to some very wide diversifications of the body. The horns, which in the wild state are invariably well developed, and which in the cattle of our Western plains attain very great size, have in certain breeds altogether disappeared, and in their place there sometimes comes a remarkable crest of bony matter which does not project beyond the skin which [114] covers the head. If such differences occurred in the wild state, they would be regarded as separating the two types of animals widely from each other.

Egyptian Sheep

In treating the wool-bearing animals along with beasts of burden, we make a somewhat fanciful classification which yet is not quite without reason. By long training man has brought these species to the state where their covering of wool or hair, once a coating only sufficient to afford protection from the weather, has become a very serious load. In certain of our highly developed varieties the annual coat is so far increased that the creature loses a large part of its bulk after the shearer has done his work. Each year's fleece often amounts in weight to eight to twelve pounds, and in its lifetime the animal may yield a mass of wool far exceeding its [115] weight of flesh and bones in any time of its life. When the fleece is mature the animal is often burdened with a load about as heavy in proportion to his size as is a horse by the weight of its rider and accoutrements.

As a flesh producer, particularly in sterile fields, sheep are more valuable than our horned cattle. They mature more rapidly, attaining their adult size and reproducing their kind in less than two years, so that in many parts of the world it is possible to obtain a larger quantity of flesh from poor pasturages with sheep than with any other of our domesticated animals. Their principal value, however, has been from the means they afforded whereby men in high latitudes have obtained warm clothing. Before the domestication of these creatures, peoples who had to endure the winter of high latitudes were forced to rely upon hides for covering—a form of clothing which is clumsy, uncleanly, and which the chase could not supply in any considerable quantity. Owing to its peculiar structure, the hair of the sheep makes the strongest and warmest covering, when rendered into cloth, which has ever been devised for the use of man. The value of this contribution is directly related to the conditions of climate. In the intertropical regions the sheep plays no part of importance. In high latitudes it is of the utmost value to man. No other of our domesticated creatures, except the camel, is so specially adapted to the needs which peculiarities of climate impose upon their possessors.

Bedouin Goat-Herd — Palestine

The relations of the goat to mankind are in certain ways peculiar. The creature has long been subjugated, probably having come into the human family before the dawn of history. It has been almost as widely disseminated, among barbarian and civilized peoples alike, as the sheep. It readily [116] cleaves to the household, and exhibits much more intelligence than the other members of our flocks and herds. It yields good milk, the flesh is edible, though in the old animals not savory, and the hair can be made to vary in a larger measure than any of our animals which are shorn. Yet this creature has never obtained the place in relation to man to which it seems entitled. Only here and there is it kept in consider [117] able numbers or made the basis of extensive industries. The reason for this seems to be that these animals cannot readily be kept in flocks in the manner of sheep. They are only partly gregarious, and tend to stray from the owner's keeping. There seems reason also to believe that they cannot easily be made to vary in other characteristics except their hairy covering at the will of the breeder, and so varieties cannot be formed, as is the case with sheep, to suit each peculiarity of soil and climate. Thus in Europe, where it would be easy to name a score of distinct breeds of sheep, each peculiarly well suited to the conditions of the country where it had been developed, the goats are singularly alike. The original stock of these creatures appears to have been adapted to feeding on the scant herbage which develops in rocky and mountainous countries. They do not seem able to make the perfect use of the resources of a pasture which sheep do. These inherited peculiarities in feeding enable them to pick up a subsistence where they may range over a considerable territory, even where it seems to afford no forms of food for the hungriest animal. Thus in that part of the city of New York known as "Shanty town," goats may be seen in fairly good condition, although the sole source of food, besides a few stray weeds, appears to be the paste of the paper advertisements which they pick from the rocks and fences.

Although goats appear to be characterized by invariable bodies, our sheep are, in physical characteristics, among the most flexible of our domesticated animals. They may by selection readily and rapidly be made to vary as regards the character of their wool, the size and proportion of their muscles, and the quantity and placing of the fat. In all these features they may be fairly blown to and fro by the wind of [118] favor. Between the meagre-bodied merino, with its skeleton-like frame and heavily wrinkled skin bearing a vast burden of long wool, and the heavy Hampshire-downs or South-downs, there is really an immense difference in bodily quality; yet these variations represent only a century or two of careful experiment on the part of the breeders. It seems not improbable that in the present state of this developing art it would be possible, in a hundred years, to reverse the conditions of these two varieties.

Sheep and goats, like the other herbivorous species which are the common tenants of our fields and forests, belong to the great class of dull-witted mammals in which the intellectual processes appear to be almost altogether limited to ancient and simple emotions, such as are inspired by fear or hunger. They are characterized by little individuality of mind, and although the needs of men have not led to any experiment in developing their wits, as in the case of dogs, there is no reason to believe that they afford much foundation for such essays. The present rapid variations in the physical characteristics of our sheep which are induced by the breeder's skill, make it evident that we are far from having attained the maximum profit from these creatures. The goats also give promise, when selective work is carefully done upon them, of giving much more than they now afford to the uses of mankind; but from neither of these forms is there reason to hope, at least on our present lines of experiment, for any considerable gain in the intellectual qualities.

The Great Caravan Road — Central Asia

We have already noted the fact that the sheep is especially adapted to serve man in high latitudes, where he has to provide against the winter's cold. The camel is an even more striking instance in which the value of the creature [119] depends upon climatal peculiarities. It is peculiarly fitted, by its ancestral training and development, for the use of men who dwell in arid countries. In the olden days of the later Tertiary epoch, creatures akin to the camels appear to have been widely distributed, and were probably adapted to considerable variations of environment. Within the time of which we know something by history, these forms have been limited to the arid districts of southwestern Asia and northern Africa. It is not certain that we know the originally wild form of either of the two species, the double-humped or single-humped camels. Wild members of each exist, but they may be the descendants of the domesticated forms. It seems probable that long before the building of the Pyramids the people of the deserts had learned how to profit from the very peculiar qualities of this strangely provided beast, which in several distinct ways is singularly fitted to serve the needs of [120] man in arid lands. The large and well-padded foot of this creature is well adapted for treading a surface unsoftened by vegetation. Its peculiar stomach enables it to store water in such a manner that it can go for days without drink. In the humps upon its back, as in natural pack-saddles, it may harvest a share of the nutriment which it obtains from occasional good pasturages, the store being laid away in the form of fat which may return to the blood when the creature would otherwise starve. So important have these peculiarities been found by men who have domesticated the camel, that on them have rested many of the most interesting features of race development in the history of our kind. In the territories along the eastern and southern shores of the Mediterranean, and in a large part of southern and central Asia, the camel has done service to man which elsewhere has been performed by sheep, cattle, and horses. In those parts of the world the share which these domesticated animals have had in the development of man has been relatively small. The camel has given the strength for burdens, hair for clothing, and often flesh to the needy men of the desert.

[121-2]

The Halt in the Desert at Night — The Story Teller

Although long a captive, and for ages, perhaps, the most serviceable of all the creatures which man has won from the wilds, the camel is still only partly domesticated, having never acquired even the small measure of affection for his master which we find in the other herbivorous animals which have been won to the service of man. The obedience which he renders is but a dull submission to inevitable toil. The intelligence which he shows is very limited, and, so far as I can judge from the accounts of those who have observed him, there is but little variation in his mental qualities. As a whole, the creature appears to be innately the dullest and [123] least improvable of all our servitors. The fact is, this animal belongs to an ancient and lowly type of mammals characterized by relatively small brains, and therefore of weak intelligence; but, for its singular serviceableness in drought-ridden countries, it would probably have been hunted off the earth by the early men, as have been many other remnants of the ancient life.

Camels Feeding

It is somewhat characteristic of the older forms of animals, those which took shape in the earlier Tertiary periods, that they are less variable than those which acquired their characteristics in times nearer our own. It is a fact well known to the students of paleontology, that species and genera [124] which have been long on the earth are apt to become in a way rigid as regards their qualities of body and mind. It is an interesting fact that, although the camel can readily be transplanted to many other parts of the world, where the physiographic conditions are similar to those of the realm where he has served man so well, he has never been thoroughly successful except in the regions where he has been in use for ages. In the desert regions of the Cordilleras of America, in South Africa, and in Australia, various experiments go to show that the creature could be perfectly reconciled to its environment. Many years ago a lot of camels were brought to the valley of the Rio Grande with a view to their utilization in that region, which closely resembles the desert countries about the Mediterranean. These animals were thoroughly successful in meeting the climatal conditions of the region. They proved as strong and as fertile as in their natural realms. Although it is said they survive to the present day, they have never been of any service to the people.

[125-6]

Carrying the Sugar Cane in Harvest — Egypt

Although, as before noted, the camel has a certain value for other purposes than conveying burdens, these subsidiary uses are so far limited that the creature is not likely to retain a place in the world after his service in caravans is no longer called for. The rapid recivilization of northern Africa, leading as it does to the development of a railway system in that region, promises to displace this creature from his most trodden ways. It seems likely that the other portions of the desert lands in the old world will soon be brought under the same civilizing influences, the nomadic tribes reduced to a stationary habit of life, and the commerce effected in the modern manner. When this change is brought about, this old-time animal, which but for the care of man would have [127] probably long since passed away, will be likely, save so far as it may be preserved through motives of scientific interest, to join the great array of vanished species.

Camels along the Sea at Twilight

It affords a pleasant contrast to turn from the consideration of the camels to a study of the elephants. The difference in the measure of attractiveness of the two forms is very great, and depends upon facts of remarkable interest. Unlike the camel—which, as we have seen, is the last survivor of an ancient lineage, represented by but two species, and these limited to a small part of the world—the elephant, at the time when man appears to have taken shape, seems to have existed on all the continental lands except Australia, and to have been in a state of singular prosperity. As is often the case with other vigorous genera of mammals, the species were adapted to a very great variety of climates, and were fitted to endure tropic heat as well as arctic cold.

The group of elephants is first known to us in the early [128] part of Tertiary time. From its first appearance on our stage it seems to have been successful in a high measure, and this probably by reason of its possession of the remarkable invention of the trunk—a prolonged and marvellously flexible nose which serves in the manner of an arm and hand for gathering food.

When we first find traces of mankind in the records of the rocks, in what appears to be an age just anterior to the Glacial epoch, the elephant had passed the experimental stages of its development and was firmly established as the king of beasts. In his adult form he had nothing to fear from any of the lower animals, and by the organization of herds it is probable that even the young were tolerably safe from assault. Until the early races of men had attained a considerable skill in the use of weapons, the great beasts were probably safe from human attack. We may well believe that primitive savages shunned them as unconquerable. As early, perhaps, as the closing stages of the Glacial epoch in Europe, we find evidences which pretty clearly show that the folk of that land, probably belonging to some race other than our own, had attained a state of the warlike arts in which they could venture to hunt this creature.

The species of elephant which was hunted by the early men of Europe, and perhaps also by those in Asia and America as well, was a greater and, at least in appearance, a more formidable monster than the living species of Asia or Africa. He was on the average taller and probably bulkier than any of his living kindred. The tusks were large and curved in a curious scimitar form. Adding to the might of its aspect was a vast covering of hair, which on the neck appears to have had the form of a mane. This covering must have greatly [129] increased the apparent size of the creature, which no doubt appeared about twice as large as any of our modern elephants which are nearly hairless. Although the perils of this ancient chase must have been great, the triumphs were equally so, and to a people who lived by hunting, most profitable; a single animal would furnish more food than scores of the lesser beasts such as the reindeer.

It seems probable that the ancient northern elephant continued in existence in North America down to the time when this continent was inhabited by man. It can hardly be doubted that the very ancient human beings, whose remains are preserved to us beneath the lava streams of California, dwelt on the continent along with the mammoth. In excavations which I have made at Big Bone Lick in

Kentucky, where a group of saline springs emerges at the bottom of a valley, there were disclosed a very great number of skeletons of this great elephant, commingled with the bones of one or two smaller forms of the related genus, the mastodon. At a slightly higher level was the multitude of remains belonging to an extinct species of bison which came just before our so-called buffalo, while near the surface of the ground was found the waste of the creatures which were in the field when it was first seen by the white men. A very careful search failed to reveal any trace of man until the uppermost level was attained. The facts, which cannot well be discussed here, have led me to the conclusion that only a few thousand years can have elapsed since the mammoth and the mastodon plentifully abounded in North America; but I am forced to doubt whether our savages were here in time to make acquaintance with these animals.

It is not certain that the extermination of the great north [130] ern elephant or mammoth even in the Old World came about through the action of man. It is possible that the death was due to more natural causes, such as the change of climate which attended the decline of the Glacial period, or to the attacks of some insect enemy like the tsetze fly of South Africa, which occasionally brings destruction to cattle in that part of the world. On the whole, however, it seems most probable that the extermination of this noble beast is to be accounted among the brutal triumphs of mankind, perhaps as the first of the long tale of destructions which he has inflicted upon his fellow-creatures. However this may be, it is clear that at the dawn of civilization the species of the genus elephas had become limited to that part of the African continent which lies south of the Sahara, and to the portion of Asia east of the Persian Gulf and south of China. The remnant consisted of two species: the African form, on the average the larger of the two, a fierce and scarcely domesticable creature; and the Asiatic, a milder-natured species which alone has been to any extent brought into the service of man.

It is not certain when or where elephants were first reduced to domestication. In the dawn of history we find them used to enhance the state of princes and for the purposes of war. It seems possible that in this early day the African as well as the Asiatic species was tamed, at least to the point where they could be made to serve in battle. We can hardly believe that all these animals which were at the command of Hannibal and the other generals of North Africa, came from the Asiatic realm. The fact that in modern times the species which dwells south of the Sahara has not been turned to the uses of man, may be accounted for by the lowly estate of the native people in that part of the world, and the lack of need [131] for such creatures in the economic conditions of the Aryan folk who have settled along the shores and in the southern part of that continent.

The relations of man to the elephant are more peculiar than those which he has formed with any other domesticated animal. Although the creature will breed in captivity, its reproduction in that state is exceptional, and it is many years before the offspring are fit for any service. It is indeed about thirty years before the creature is sufficiently adult to attain a good measure of strength and endurance. It has therefore been the habit of the people who avail themselves of this admirable beast to use the captures which they make in the wilderness. It is a most interesting and exceptional fact that these captive elephants, though bred in perfect freedom and provided with none of those inherited instincts so essentially a part of the value of our other domesticated quadrupeds, become helpful to man and attached to him in a way which is characteristic of none other of our ancient companions except the dog. It is safe to say that the Asiatic elephant is the most innately domesticable, and the best fitted by nature for companionship with man, of all our great quadrupeds. The qualities of mind which in our other domesticated quadrupeds have been slowly developed by thousands of years of selection and intercourse with our kind, are in this creature a part of its wild estate.

It appears from trustworthy anecdotes that the Asiatic elephants in a few months of captivity acquire the rules of conduct which it is necessary to impose upon them. The speediness of this intellectual subjugation may be judged from the fact that, after a short term of domestication, they will take a willing and intelligent part in capturing their kindred [132] of the wilderness, showing in this work little or no disposition to rejoin the wild herds. In the case of no other animal do we find anything like such an immediate adhesion to the ways of civilization. We have to account for this eminent peculiarity of the elephant on the supposition, which appears to be thoroughly justified, that the creature has, even in its wild state, a type of intelligence and instincts more nearly like those of men than is the case with any other wild mammal, an affinity with human quality which is, perhaps, only approached by certain species of birds. It appears from the observations of naturalists that the family or tribe of wild elephants is a distinct and highly sympathetic community. The grade and value of the friendly feeling which prevails among them may be judged by the fact that, when one of the males becomes lost or is driven away from its associates, it does not seem to be able to join any other tribe, but becomes a "rogue," or solitary individual, and in this state develops a morose and furious temper.

There are many well-attested stories which serve to show that wild elephants have a kind of intelligence which indicates a certain constructive capacity. Of these, perhaps the best are the instances in which the creatures have been caught in pitfalls, made by digging a hole in the paths of the wilderness which they are accustomed to follow, the surface being covered with a frail platform so arranged as to conceal the excavation. When one of a tribe is caught in the trap, the others, if time allows before the hunters come to the ground, will in an ingenious way release him. I doubt if the most practicable manner of effecting this will occur at once to the reader. The easiest plan may seem to drag the captive from the pit by sheer strength, but as the hole is deep and has vertical sides, [133] the elephants contrive a better way. They bring bits of timber, which they throw into the pitfall, the captive treads them down until he is elevated to a position whence he can escape from his prison.

The intelligence of the wild elephant is probably in good part to be accounted for by the fact that the creature possesses in its trunk an instrument which is admirably contrived to execute the behests of an intelligent will. It is easy for us to see how, in the case of man, the hands have served to develop the intelligence by providing him with means whereby he could do a great variety of things which demanded thought and afforded education. The elephant is the only large mammal which has ever acquired a serviceable addition to the body such as the trunk affords. In their ordinary life the trunk does almost as varied work as the human arm. With it they can express emotions in a remarkable way; they caress their young, gather their food by a great variety of movements, or defend themselves from assailants. To the naturalist who has come to perceive the close relations between bodily structure and mental endowments, it is not surprising to find that these creatures have attained a quality of mind which is found nowhere else among the mammals except in man and in some of his kindred, the apes.

The most peculiar mental quality of the elephant, a feature which separates him even from the dog, is the rational way in which he will do certain kinds of mechanical work. He appears to have an immediate sense as to the effects of his actions, which we find elsewhere only among human beings. From a great body of well-attested observations, showing what may be called the logical quality of the mind of these creatures, I may be allowed to select a few stories which have [134] a singular denotative value. An acquaintance of mine, a British officer who had served long in India, told me that in taking artillery over very difficult roads, certain of the abler elephants could be trusted to walk behind each piece, where they would in a fashion control its movements, steadying or lifting it as the occasion demanded without any directions from the driver.

An Indian Elephant

Elephants can be trained to pile up sticks of timber, such as railway ties, placing the layers alternately in opposite directions, as is the custom in such work. There is an excellent and well-attested story of an elephant who, without a driver, was bearing a stick of timber through a narrow wood [135] path. Meeting a man on horseback, and perceiving that the way was not wide enough for both himself and the oncomer, the sagacious animal deliberately backed his huge body into the chaparral so as to clear the way, and then trumpeted as if to signal the horseman that the path was free.

The emotions as well as the intelligence of elephants are singularly like those of human kind. It is said by those who know them well that if when in their stubborn fits they are brutally overborne, they are apt to die of what seems to be pure chagrin. Their states of grief, despair, and rage much resemble those which are exhibited by violent children or men unaccustomed to control. Their affections and animosities have also a curious human cast. They readily form attachments which appear to be quite as enduring as those exhibited by dogs, and their memory of injuries remains quick for years after they have received the harm. Well-verified anecdotes showing the likeness of these emotional qualities to our own exist in such numbers that it would be easy to fill a volume with them. They are, however, not necessary to show the likeness of the creature to ourselves. This is sufficiently exhibited by their daily behavior under domestication. In noting this we should remember that the male elephant is the only large mammal the males of which it has proved safe to use in the ordinary work of life. Even our bulls and stallions, though they belong to species which have been domesticated for thousands of years, are so violent and untrustworthy as to be of little value except for breeding purposes. Bulls, even of the tamer breeds, are a constant menace to the lives of their masters; yet an adult male elephant recently made captive may, except when seriously diseased, be trusted to obey the mere signals of the driver, who has no such control [136] over him as the bit affords in the case of horses. The creature has the strength to overcome all control save that of a moral nature. To this he submits in a way which is only equalled by our well-bred dogs.

As yet the utility of the elephant to man has, measured by his qualities, been but small. The creature has a marvellous strength, great intelligence, and remarkable docility. In proportion to the power which he can apply to a task, he is not an expensive animal to maintain. He can endure a considerable range of climate, and enjoys a tolerable immunity from disease. The reason for the relatively inconsiderable use of these creatures is probably to be found in the fact that they are not adapted for ordinary draught purposes, nor are they well suited to the needs of the caravan, for which the camel or the pack-mule is much better fitted. In ancient warfare, before the invention of gunpowder, elephants carrying archers or javelin-men upon their backs were greatly valued for the effect of their charge against an enemy and for the fright with which they inspired horses. Against the unsteady ranks of Oriental armies they were often most efficient in breaking a line of battle. Even the Roman troops, when they first encountered them and before they knew how to meet their charges, found them very formidable. It was soon learned that if their onset was stoutly resisted, they were likely to become unmanageable in the uproar of the fight, and to do as much damage to friends as to foes. It is only in certain peculiar tasks that, in modern days, the elephants have any economic value, and in the most of this work their strength is likely to be replaced by various engines.

The two existing species of elephants are, as before remarked, the survivors of a long lineage, represented in the [137] geological record by the remains of many extinct forms. Some of these lost species were far smaller than those of to-day; one at least was no larger than our heavier horses. If by the breeder's art the existing varieties could be caused so to change as to give us once again this relatively diminutive form, the creature would be sure to find a place of importance in our ordinary arts. The trouble is that the very long life of this animal is naturally associated with a slow growth. It requires indeed almost the lifetime of a generation to bring the individual to an adult age. It is therefore not surprising that, as the wild forms can readily be won to domestication, these creatures have not been the subject of any of those interesting processes of selection which have so far affected for the better the characteristics of nearly all the other domesticated animals.

In every other regard than those mentioned above, the elephant appears to be an excellent subject for improvement by choice in breeding. The individuals vary much as regards their physical and mental qualities. Probably no other wild mammal exhibits such differences in the mental features as does this highly intellectual creature. The physical individuality does not seem to be as striking as the mental, but even here we note a range, at least as regards size, which is unusual in the wild forms bred under similar conditions. The general elasticity of the group is shown by the considerable differences which may be traced in the herds which occupy different parts of the field over which the species range. As yet these local peculiarities have not been carefully studied; but from an examination of the tusks in the ivory warehouse at the docks in London, I have found that those shipped from particular ports in Africa and Asia differed both in form and [138] texture, so that the experts were able to tell from which district they came. The evidence, in a word, appears to show that the creature tends to vary; and it is a safe presumption that the forms would prove as responsive to the breeder's art as those of our horses, cattle, sheep, or dogs.

As a whole, the elephant has been almost as little associated with the life of our own race as the camel. Neither of these creatures has ever played any considerable part in European affairs. From the disappearance of the last of the mammoths in the closing stages of the Glacial time until the invasions of Italy by Pyrrhus and by Hannibal, elephants were practically unknown in Western Europe. They have never been used in peaceful occupations on that continent, and have had only a trifling place in its military arts. It was probably due to this separation of our eminently experimental race from the realm of the elephants that no efforts have been made systematically to breed them in captivity, and thus to win varieties in which the form might become better adapted to economic needs, and the remarkable mental powers of the creature be brought to their utmost development. As yet the only Europeans who have had much to do with elephants are the British, who in their civil and military service in India have been thrown in contact with these animals. Generally, however, these people have been only temporarily domiciled in Asia, and probably on this account have not become interested in the problems which this noble beast presents to all those who appreciate the animal world. We lack, indeed, the observations which might have been made with admirable effect by British observers in India during the two centuries in which that people has had to do with the lands in which elephants abound.

[139]

The elephant of Africa is still a tolerably abundant animal. Its numbers, though doubtless diminished by more than one-half within this century, are probably to be counted by the hundred thousand. Nevertheless, in less than a hundred years the field which they occupied has been greatly reduced; and between the ivory hunter and the sportsman of our brutal race armed with guns of ever-increasing deadliness, it will certainly not require another century of free shooting to annihilate the African species. In view of the present condition of the life of these noble beasts, it seems in a high measure desirable that a thorough-going effort should be made to extend the domestication to the point where the form will not only be won from the wilds, but will be a permanent element in our civilization, in the manner of our common flocks and herds. It will be an enduring shame if, by neglect of our opportunities, the utmost is not done to attain this end. It appears fit that this task should be undertaken by the British Government, which in modern days has displayed a skill and forethought in the administration of its Indian provinces unexampled in the history of colonies. Owing to the slow breeding-rate of the elephant, it may require more than a century for experiments to attain any definite result, so that the task is clearly beyond the limits of individual endeavor.

Among the humbler helpers of man, the pig holds an important place. He has had no small share in the betterment of the estate of his masters. One of the large questions which beset men in their unconscious endeavors to lay the foundations of civilization was that of food-supply. No sooner does a population become sedentary than the wildernesses about its dwelling-place are rapidly cleared of the large [140] game, so that the chase affords but little save amusement. Therefore a provision in the way of meat has to be obtained from domesticated animals. The flocks and herds supply this need, though in a costly way. Sheep have a value for their wool; horned cattle develop slowly, and are, moreover valuable, the oxen for their strength and the cows for their milk. Horses are too valuable to be used for food, save in times of exceeding stress; and none but the lowest savages are willing to send their faithful dogs to the pot. From the beginning of his experience with man the pig has been found the cheapest and most serviceable domesticated animal as a source of food-supply.

We can trace the origin of our domesticated pigs more clearly than in the case of the most of the other subjugated animals. The creature is evidently descended from the wild boar of Europe and Asia; and though long under domestication and greatly varied from its primitive stock, it readily reverts to something like its original form when allowed to betake itself once more to the wilds. The domestication of the species appears to have been accomplished at several different points in Asia and Europe. The forms which are found in eastern Asia differ from those which are kept in the western portion of the great continent, and may have their blood commingled with that of another species which is native in that part of the world.

Among our domesticated animals the pig is exceptional in the fact that it has been bred for its flesh alone; for although the hide is valuable and the hair serves certain purposes, as in the manufacture of brushes, these uses are only incidental and modern. They have not affected the plan of the breeder, whose aim has been to produce the largest weight of flesh in [141] the shortest time, and with the least expenditure of food. In this peculiar task the success has been remarkable, the creature having been made to vary from its primitive condition in an extraordinary manner. In its wild state the species develops slowly, requiring, perhaps, three or four years to attain its maximum size. It never becomes very fat, but remains an agile, swift-footed, and fierce tenant of the wilds. Under the conditions of subjugation the pig has been brought to a state in which its qualities of mind and body have undergone a very great change. In the more developed breeds, even the males, when kept about the barnyard, are quiet-natured and not at all dangerous. The creatures have become slow-moving; they attain their full development in about half the time required for the growth of their wild kindred, and when adult they may outweigh them in the ratio of four to one.

The effect arising from the food-supply which our pigs afford is well seen in the use which is made of their flesh in all the ruder work of men, at least in the case of those of our race. Our soldiers and sailors are to a great extent fed on the flesh of these creatures, which lends itself readily to preservation by the use of salt. So rapidly can these animals be bred, owing to the number of young which they produce in a litter and the swiftness of their growth, that sudden demands for an increase in the supply, such as occurred at the outbreak of our civil war, can quickly be met. If the need should arise, the quantity of pork produced in this country could readily be doubled within eighteen months. This is the case with no other source of flesh-supply, and this fact gives the pig a peculiar importance.

Owing to the remarkably complete domestication of this [142] animal, and also to the fact that it is omnivorous, the creature has ever been a favorite with the cotter class. Those folk, who can afford neither sheep nor horned cattle, can often provide the food for pigs, and thus, in turn, be much better fed than they would otherwise be.

It is only within two centuries that our pigs have attained to anything like the domestication in which we commonly find them. Of old they were allowed to range the forests, much as they do in certain parts of our Southern States at the present day. In some parts of Europe, particularly in the southern portion of the continent, this method of rearing and feeding is still common. It was and is advantageous, for the reason that the creature, by its remarkably keen sense of smelling and its singular capacity for overturning the ground, is able to provide itself with abundant food in the way of grubs and roots which are not at the disposition of any other animal. It was only as the public forests disappeared that pigs came to receive any considerable part of their provender from the products of tilled fields. In this stage of our agriculture, when all the land was possessed, the life of the pig was necessarily more restricted, and he became the denizen of a pen. In the earlier state there was no cost for his keeping; in the latter, except so far as he could be fed from the waste of a household, he is an expensive animal.

It is with this last state of the pig, when he became the most housed of our domesticated animals, that the work of the breeder really began. The aim of those who have developed the pig has been, as we have said, to obtain the most rapid growth along with the greatest weight of fat, and to accomplish the results with the least expenditure in the way of food. Although the animal has been subjected to selective experi [143] ments, looking to these ends, for not more than a century, or say about forty generations of the species, the amount of variation which has been attained is singularly great, the form and habits having been changed more rapidly, and in a larger measure, than in the case of any other of our domesticated animals. It may fairly be said that this creature is more obedient to the will of the practical selectionist than any other with which we have experimented.

It is commonly assumed that our pigs are among the least intelligent of the creatures which man has turned to his use. This impression is due to the fact that the conditions in which these animals are kept insure their degradation by cutting them off from all the natural mental training which wild animals, as well as the other tenants of the fields, receive. In the state of nature or in the condition of domestication which existed before pigs became captives in their pens, they were among the most alert and sagacious animals with which man has come in contact. Their wits were quick and their sympathies with their kind remarkably strong. Trainers have found these creatures more apt in receiving instruction than any other of our mammals, and the things which they can be made to do appear to indicate a native intelligence nearer to that of man than is found in any other species below the level of the apes.

As there is little in the books of anecdotes of animals concerning pigs, I venture to give an account of a learned individual of this species whose performances I had an opportunity of observing in much detail. The creature, an ordinary specimen about three years old, had been trained by a peasant in the mountain district of Virginia who made his living by instructing animals for show purposes. He stated [144] that in selecting pigs for education it was his practice to choose those characterized by a considerable width between the eyes and whose skulls projected in this part of their periphery to a more than usual degree. He said that from many experiments he was satisfied that there was a very great difference in the capacity of the animals to receive training, and that the above-mentioned indices afforded him sufficient guidance in his choice.

In the exhibition about to be described there were but three persons present, myself, another spectator, and the showman. A score of cards were placed upon the ground, each bearing a numeral or the name of some distinguished person. These cards were in perfect disorder. I was allowed, indeed, repeatedly to change their position and to mix them up as I pleased. The pig was then told to pick out the name of Abraham Lincoln and bring it to his master. This he readily did. He was asked in what year Lincoln was assassinated. He slowly but without correction brought one by one the appropriate numerals and put them on the ground in due order. Half a dozen other questions concerning names and dates were answered in a similar way. Each success was rewarded with a grain of corn, and for his failures the creature received a reasonable drubbing. It was evident that the animal had to consider in making his choice of the cards. At times he was evidently much puzzled and would indicate his perplexity by squealing.

It seemed clear that the master of this learned pig did not guide the movements of the animal by other indications than words. The questions, in some cases, had to be reiterated in a loud voice in order to insure attention. Several times during the performance the pig rebelled, broke from the tent, [145] and was with difficulty recaptured. The creature disliked this task in the manner of a lazy school-boy, and at the end of an hour of exercises seemed utterly overcome by his labor. He ran into the box where he was ordinarily confined, and when dragged forth, neither rewards nor punishments would quicken him to further work.

The above-described exhibition made it plain to me that the pig can be taught to understand a certain amount of human speech and to associate memories with phrases substantially as we do ourselves. It is perfectly clear that the performance which I witnessed was not a mere routine action, for I had a number of questions asked over again so as to make it sure that the creature acted with reference to each separate inquiry. The behavior of the animal during the performance seemed clearly to indicate mental effort and not mere automatic memory. His attitude when trying to determine which of two cards to take distinctly showed that he was intently viewing the figures and endeavoring to come to a decision. I am aware it has been suggested that learned pigs discriminate between the cards by peculiarities of odor which have been given to these bits of paper. I sought carefully to find if such was the case, and though I have a very keen sense of smell I found nothing which led me to suspect that this device was used. Even if such were the case, the rationality of the animal's action would be none the less clear. The showman assured me that he never used any such means in training pigs. He seemed, indeed, to treat the suggestion with contempt.

Although experiments in the training of pigs show that they have rather remarkable intellectual capacities, the most human feature in their mental organization is found in the [146] keen sympathy which they exhibit with the sufferings of their own kind and the willingness with which they encounter danger in protecting their comrades. It usually requires close observation for the naturalist to determine the existence of this motive among the other wild or domesticated mammals. In fact, the traces of it are very slight indeed, and are generally to be attributed to the care of parents for offspring or of the males for their harem—a disposition which, though akin to the defence of the kind, is nevertheless of a special and peculiar nature. Even among our domestic dogs, whose sympathies have been developed in a remarkable degree and who will sacrifice their lives to defend or rescue the human beings with whom they are familiar, there appears to be but little disposition to support members of their species who may be assailed. With pigs, however, as is well known to all those who have observed their habits, the characteristic cry of distress of their fellows proves very exciting and stimulates all the adults, both male and female, who hear it to hasten in defence of their kinsmen. It is a noteworthy fact that while most other animals when in danger utter no distinct or continuous cry, the pig gives voice in a vociferous and insistent manner, as if he had a right to expect the sympathy and help of his species. The cry goes with the custom of defence which in this species has attained a better foundation in the sympathetic motives than in any other mammal below the level of man.

It is perhaps due to their relatively high intellectual organization that the excessively domesticated pigs are liable to suffer from attacks of mania. This is most commonly exhibited by the sows, which at times will destroy their young shortly after they are born. The sight of their [147] progeny seems to infuriate them in a curious manner. One sow which I owned killed three successive litters; another fine animal of the Berkshire breed, a very amiable, indeed affectionate, creature, was carefully watched at the time she first bore young, precautions being taken to prevent her from harming them; she would willingly allow them to suckle, provided she did not see them, but the moment she laid her eyes upon them she was seized with the strange fury.

Although this singular perversion of the natural instincts of maternity sometimes occurs among the pigs which are allowed to roam together in herds, it seems to be far more common in those conditions where the animals are confined in pens without contact with their kind, and where they have no chance to recognize the young as members of their species or to acquire that interest in them which they would gain in the society of the herd. It is also clear that this maniacal habit is inherited; according to my observation it is common among the Berkshire, and relatively rare in other less specialized varieties.

The intelligence of the pig is also shown in the readiness with which the creature changes its habits to meet varied environments. Thus the pigs which range the woods in the western and southern parts of the United States have learned to catch the crawfish which abounds in the shallow streams in those parts of this country. They will wade up a brook, turning over the stones and driftwood as they go, catching with a quick movement the crustaceans which they have thus dislodged from their cover. Along the shores of the Bay of Fundy, the pigs, accustomed to follow the tide out, picking the chance food which is thus exposed to them, have learned carefully to avoid the risk of being caught by the returning [148] waters. With the first splash of the turning tide they hasten inshore until they have attained safe ground.

One of the best evidences of the mental state of these animals is found in their actions when assailed by dogs or other beasts of prey. Pigs, though wary and sensible of danger, seem exempt from the extreme fear which leads to panic, and fight, even before being brought to bay by long chasing, in a discreet and valiant manner. Where a number of them are attacked by dogs or other enemies, they will form a circle with their heads out, each supporting the other in such a manner that the ring cannot readily be broken. Their thick-skinned forequarters and stout tusks provide them with excellent instruments with which to resist an assault.

The sagacity of the pigs is probably, in part at least, to be attributed to the fact that in their native state they are communal animals, all the species of their family being accustomed to live gregariously, so that for ages they have had the training which every social organization, however simple, affords. They are, moreover, omnivorous feeders, accustomed to subsist on a great variety of food—a habit which seems in all cases to promote the development of the intelligence in animals.

Although the pigs by their nature afforded the best opportunity for developing an intellectual animal which has come to us through our domesticated creatures, no effort whatever has been made by selection to develop the latent mental capacities of this species. It is perhaps the only form of those which man has subjugated which by his treatment he tends to degrade. In the time to come, when men will be held to a better accountability for the treatment of their [149] captives, the condition of these animals will afford a fair field for the reformer's care.

The geologist who is acquainted with the mammalian life of the Middle Tertiary period readily notes the fact that the variety in genera and species appears to be much greater than it is at the present time. A great number of forms, differing somewhat widely from those now in existence, then abounded in the Americas and the Old World. It may at first sight seem unfortunate that man did not have the chance to essay his domesticative arts on that older and apparently richer life. A closer examination, however, leads us to see that the species of that time, though more numerous than those of the present, were on the whole less fitted for our use than the fewer but more completely differentiated kinds with which we have had to deal. The multitude of kinds which we find in the Mesozoic period indicates that the life was in a state more experimental than that to which it has attained. A host of forms on their way towards the specialization which has now been attained have been removed from the sphere, in the manner of a scaffolding from a completed structure. That which has been left remains because it has successfully accomplished the task of reconciliation with environment, or, in simpler phrase, because it has learned to do things which were useful and profitable in a more perfect manner.

As an illustration of the fact that the animals of to-day are better fitted to be the help-meets of man than were their ancestors of an earlier time, we may note the state of the horse at the time when that genus was undergoing its development in the region about the upper waters of the Missouri. As may be imagined, the long and difficult passage from the [150] five-toed to the single-toed form was slowly accomplished, and to its doing went a great many temporary forms, which served, we may say, as stepping-stones for the ongoing. So far as we can judge, these intermediate forms were small, rather frail creatures, which probably could not have been made to serve any purpose useful to man. It was not until the mechanical system of the large single toe with the wonderfully developed nail, which makes up the foot and hoof of the horse, had been attained, that the creature becomes fit for the wonderful work we have persuaded him to do in our civilization.

A comparison of the skulls of the Tertiary mammals and those of our own day indicates that in certain of the important series, and presumably in them all, the brain has increased in size from the earlier to the later times. This increase in brain capacity has doubtless been attended by a decided gain in the measure of intelligence, a gain which has doubtless served to make the modern representatives of the series fitter for man's use than their ancestors were. For, while the number of our very useful domesticated forms may seem at first sight to be dull of wit, none of them are really low in the intellectual scale as we apply it to the brute; in fact, a considerable measure of intelligence is absolutely required as a condition for true subjugation. This is seen by the fact that nothing like a real adoption into our social system has ever been accomplished except with a few of the higher orders of mammals and birds, species which have an intellectual capacity that we recognize as akin to our own. Thus, so far as we can see, man's appearance on this stage was, so far as it relates to the possibility of companionship with the lower life, exceedingly well timed. He came at a period when the life was ready to give him and to receive from him a large [151] measure of help. If his advent had been much earlier, he might have had less trouble in his contests with the larger carnivora; but if there had been a lack of beasts to obey his will, it is doubtful whether he could himself have won his way above that primitive life.

[152]

DOMESTICATED BIRDS

Domestication of Animals mainly accomplished by the Aryan Race; Small Amount of Such Work by American Indians.—Barnyard Fowl: Mental Qualities; Habits of Combat.—Peacocks: their Limited Domestication.—Turkeys: their Origin; tending to revert to the Savage State.—Water Fowl: Limited Number of Species domesticated; Intellectual Qualities of this Group.—The Pigeon: Origin and History of Group; Marvels of Breeding.—Song Birds.—Hawks and Hawking.—Sympathetic Motive of Birds: their Æsthetic Sense; their Capacity for Enjoyment.

It is an interesting fact that about all the work of domestication which has been done by man has been accomplished by the peoples of Asia and mainly by the Aryan race. The American Indians tamed the llama and alpaca and a few species of native plants; even where their habits were prevailingly sedentary they domesticated no birds. It was left for Europeans to make use of the wild turkey. Our primitive people had the same chance to tame ducks and geese as the folk of the Old World. They appear, however, to have lacked all capacity for such endeavors. The same lack of disposition to capture and tame wild creatures is noticeable among the characteristic peoples of Africa; all of which serves to show that the domesticating art, at least as applied to animals, is peculiar to the higher-grade folk of the Old World.

Of all the birds which have been domesticated, our common barnyard fowl has been by far the most useful to man. It has become in a way interwoven with his life to a degree found only in a few of our barnyard animals. Next after the pigeons and the pigs it has been most deeply impressed by [153] the breeder's art. The wild species whence it sprang is a small creature, laying but few eggs and with but a slight tendency to accumulate fat. From this parent stock varieties have been bred which attain in some cases to eight or ten times the weight of the ancient form. They have, moreover, lost the fierce combative spirit which characterizes their ancestors and which by selection has been preserved and intensified in our breeds of game-cocks.

The Original Jungle Fowl (*Gallus bankiva*)
and Some of His Domestic Descendants

It is an interesting fact that our barnyard fowl is the only species of a large family of birds which has been truly domesticated. The kindred pheasants and grouse, though abounding in the Old World and the New, and much disposed to abide about the cultivated fields, appear to be rather untamable. However well cared for, the wilderness motive seems never to have been eradicated. The domesticability of the cock, as is that of most other wild animals, is doubtless to be explained by the conditions of the life in which it has dwelt for ages before it was introduced to the society of man. In its [154] wild state this bird had already to a great extent lost the power of flight, using its wings only for escaping from four-footed pursuers or to attain the branches of the trees in which it sought safety in the night time. With this measure of loss of the flying power, the creature abandoned the habit of ranging over a wide field, and thus was made more fit for domestication. Moreover, in their wilderness life these birds dwelt in more established communities than their kindred species. The most of these wild forms do not keep together through the year, but scatter after the young are able to shift for themselves. The Indian species of *Gallus*, however, from which our cocks and hens descend, have organized their life so that the individuals remain associate in a friendly way throughout the year.

A part of the fitness of this creature to cast in its lot with man arises from the fact that they have very sympathetic natures. This is shown by the way in which the cocks will fight for their hens, even against their dreaded enemies, the hawks; and by the manner in which the mother, overcoming her natural fears, will do battle for her brood. It is shown also in the curious mingling of gallantry and kindliness with which the cock will call a hen to give her some choice bit of food which he has captured. As he grows older and becomes Philistinish, we may note that, after the manner of unfeathered bipeds, he is often disposed to indulge his selfishness, and summons his flock only to see him devour the morsel. Even in old age, however, the males of the varieties which are nearest the parent stock maintain their helpful motives and will struggle with infirmity to beat off a bird of prey.

The sympathetic and affectionate quality of our barnyard fowl is perhaps best indicated by the singular variety and [155] denotative value of their various calls and cries. Those who know these birds well will find no difficulty in recognizing about a score of diverse sounds, each of which indicates a particular turn of their mind. Almost all of these different notes have slight variations of expression which fit particular situations. Thus the crow of these birds, which may seem to the unobservant a very unvaried sound, discloses to those who have lovingly studied them at least half a dozen distinct modifications. In the fledgling male who just begins to feel the spirit of his kind, and who goes through his performance in the adolescent way, it is a cheap and often pitiful call. From the open roost in the trees, where the birds are gradually aroused by the slow-coming day, we can often hear the note of the half-awakened cock, as full of the sense of slumber as the speech of a sleeping man. As the creature gradually awakens, his cry becomes more resonant until it has the true morning ring. Brave as is this note of the full day, it is not to be compared with the crowing of a game-cock, the most splendid braggart sound of all the animal world.

The really sympathetic notes of our fowls are uttered in their ordinary intercourse. Here the gradations of sounds have a range and fineness which, it seems to me, we can observe in no other creature below the level of man. Attention, astonishment, fear, commonplace distress, exultation, and agony are all set forth with cries which we, in a way, recognize as appropriate. Although some of these sounds relate to the larger experiences of the creatures, the most instructive of them are uttered in their ordinary intercourse, where they clearly maintain a kind of consensus in the flock by unending small bits of emotional speech, the notes being shaded in a wonderful way. These fine variations of utterance can some [156] times be observed to be related to slight differences of situation. Thus the cackle of a hen when she leaves her nest after laying an egg is quite different from that which is made by the same hen when, during the period of incubation, she quits her eggs in search of food and water.

It is not unlikely that the eminent domesticability of our common fowls is in a way associated with the singular variety of their notes. This variety indicates that the creatures are in constant and effective communication with one another; in a word, they are very sympathetic. With this intellectual helpfulness naturally goes the love of the domicile and a disposition to submit to control.

So nice and well understood are the differences between the sounds which these birds give forth, and so well are their notes appreciated by their companions, that the creatures may well be said to have a language. Though it probably conveys only emotions and not distinct thoughts, it still must be regarded as a certain kind of speech. The modes of expression indicate that in this creature, as in the other feathered forms, the intellectual life consists largely in the movements inspired by the emotions. On the rational side our fowls seem weaker than many other less interesting species. In their nesting and other habits there are no evidences of constructive ingenuity; and in all my observations on them I have never seen any evidence which showed either considerable powers of memory or a capacity to act in any complicated way with reference to an end. It is evident, however, that they make a very good classification of the world about them. They have, for the limited field over which they roam, a keen topographic sense; they never are lost, and this in connection with their sympathetic [157] homing instinct prevents them from wandering from their accustomed places to take up again with a wilderness life.

In their adhesion to domestication our common fowls differ in a remarkable way from all other of our captive animals except the dog, and these birds are even more ineradicably attached to man than their older companion. While the dog will sometimes become half wild, or, as we may phrase it, undomiciled, fowls seem incapable of maintaining themselves apart from human care. In much ranging of the wilderness I have never found one of these creatures more than a thousand feet away from a human habitation. When we consider how common must be the chances of their going astray, and how easy it is in many parts of the country, as in our Southern States, for them to obtain in the wilderness food throughout the year, the fact that they never go wild is indeed remarkable. It can only be explained by the great development of the homing instinct which man has brought about in their sympathetic souls.

Houdin   Cochins   Leghorns   Game

Although our unnatural process of breeding has done much to degrade the original beauty of the cocks and hens, destroying the delicate coloration of the feathers as well as the admirable blending and contrasts of their pristine hues, it seems likely that the effect on the physical and mental development as a whole has not been unfavorable. Though less courageous, they are stronger creatures than in their wild state; they are clearly more fecund; they are gentler natured; and, so far as I have been able to compare the high-bred with the primitive forms, their range of expression through the voice has been much increased, a feature which may be noted in other domesticated species of birds, as, for instance, in the canaries. The most remarkable alteration which has [158] been brought about in the minds of these creatures consists in the very great diminution in the combative motive of the males. In the wild forms, as well as in the kindred variety of the game-cock, this impulse to battle attains a truly phenomenal development, the like of which is probably not to be found in any other creature. The male birds begin their warfare before they are more than half grown, and in their adult state will attack anything which they can conceive to be an enemy. They will, with slight provocation, assail any of the other domesticated species of birds, and even the lesser mammals, such as the dogs and cats. They will fight [159] their own image in a looking-glass. I have had game-cocks attack my hand when it was held near the ground and given an up-and-down movement in imitation of their antagonist's head.

I once reared a game-cock by hand, keeping him secluded from his kind until he was adult. I then placed him in a large collection of barnyard fowl where there were half a dozen mongrel cocks, a drake of the muscovy variety, several ganders, and two turkey-gobblers. Immediately and in rapid succession he settled his accounts with the males of his own kind. He shortly overcame the drake and the ganders. He then devoted what was left of his forces to battles with the turkeys. Here he found himself in great difficulty, for the reason that these great birds would seize him by the head and lift his body off the ground. However, he soon learned an ingenious trick which protected him from this danger. When gathering breath in the intervals between his assaults, he would hover himself between his antagonist's legs, keeping step with the awkward creature in its efforts to get away from him. In a few days he wore out these doughty foemen and remained the battered master of the field.

Although the indomitable valor of the game-cock may be in some measure due to the selection which the breeder has applied to the variety, there can be no question that it is essentially natural to the species and is the result of an age-long habit which in the native wilds of the creature did much to insure its safety. The antiquity of the state of mind may be judged by the perfection to which the spurs have attained and the remarkably skilful and definite way in which the creatures use them. The spur, which has arisen from the development of the scales and underlying bone of the bird's [160] leg, is a singularly perfect structure, the finish of which cannot be judged in the degraded form in which it is found in our ordinary barnyard species. Although in its construction this weapon is admirably devised, it is placed in a position where only a remarkably well-addressed movement can give effect to its blow. Those who have watched game-cocks in combat have had a chance to see the vaults by which the creature, partly turning in the air, is able to throw the spur in such a manner that it shares the impulse of the body when it strikes the antagonist. This peculiar craft has been in good part lost among our common varieties. Their spiritless contests differ as much from those of the game-birds as do the fist fights of untrained men from the contests of skilled pugilists.

Bantams   Brahma   Dorkings

Although to persons unaccustomed to the spectacle the combats between game-birds may seem disgusting, almost [161] every one must admire the valor, grace, and address which such scenes exhibit. Except where the brutal custom of putting steel points on the spurs prevails, the birds rarely receive fatal wounds. The defeated cock is soon brought to confess his inferiority and takes himself away. At no other time in the life of these birds does their organic beauty appear to such advantage as when they are struggling with each other. Then alone do we perceive the singular efficiency of their bodies and the quick as well as appropriate action of their instincts. They set themselves against each other in attitudes as well chosen and as peculiar as those of a well-trained fencer. Before the assault they often go through a singular performance, which consists in picking up bits of twigs or pebbles. These they cast into the air, an unmeaning movement which may be compared to the like meaningless though similarly graceful salute with which swordsmen preface their contests. Then, with their legs flexed so that they may be ready for the spring, and with the rather stiff feathers about the neck erected so as to serve as a shield, they creep toward each other until they are separated by the distance appropriate for the spring. When fairly placed for battle they begin a system of fence which is intended to provoke the enemy to an untimely assault. The art of the game appears to consist in persuading the adversary to venture an attack where his force will be spent in the air, so that a blow can be given him before he has time to recover position. The issue depends much on the endurance of the birds. Their movements require so much energy that one of them is apt to become exhausted before the other is quite spent. In rare cases, only one of which has been seen by me, a weary bird will feign death for a minute or so and thus obtain new [162] strength with which to renew the combat, profiting also by the confusion which he will bring upon his adversary by his sudden revival.

Although the combatant motive which we find in the males among our barnyard fowls has doubtless been developed through their combats with each other, the valiant spirit which has come from it often leads the creatures to attack the enemies of their flock. I have seen a nimble game-cock strike a hawk which was pouncing to its prey, delivering the blow some feet above the surface of the ground, and this so effectively that the marauder was driven away in a sorely hurt condition. I have seen males of the game variety attack a number of other larger animals which in any way threatened their charges.

Although our barnyard fowl are almost the only ground birds which have ever been brought to a state of perfect domestication, there are several other species of the same group which have been taught in a measure to adhere to man. Of these perhaps the longest in domestication is the peafowl. This creature, though it has edible, indeed we may say savory flesh, has retained its small place in civilization solely on account of its extraordinary beauty. For its size it is doubtless the most beautiful of animals, its plumage, especially the magnificent display of the tail, exceeding that of any other natural object. There are other birds of small size which vie with the peacock in the details of ornamentation. Those jewels among the feathered tribes, the humming-birds, have a more delicate beauty. The birds-of-paradise and the lyre-birds have a grace in the attitudes of particular feathers which is unequalled; but for splendor none of them approach the peacock in his best estate.

Contributions from Asia, Africa, and America—Peacocks, Guinea-fowl, and Turkey

[163] The peacock is a native of Southern Asia, a realm in fact in which the species of the group attain an uncommonly rich development. The creature appears to have been domesticated some thousands of years ago, but has undergone no considerable changes in its experience with man. It has in truth not been completely tamed. It does not willingly remain near the dwellings of man, but prefers to abide apart, only resorting to the home when in [164]

need of food. It is very intolerant of the other barnyard creatures, and often becomes possessed of a kind of mania for slaying their young, not for food but from pure spirit of mischief.

Intellectually speaking, the peacocks are much below the cocks and hens; although they flock together, their sympathies do not seem quick; their cries and calls do not number a fifth part of those which we hear from our chickens, and their notes are prevailingly very discordant. Their cry of defiance, answering to the crow of the cock, is one of the rudest and least sympathetic sounds which is heard among the birds. Its only merit is that it can be heard very far. It is readily audible at the distance of a mile when it breaks the stillness of a summer night. At present the bird seems out of favor. At best it is a beautiful but annoying ornament to pleasure-grounds. It is likely, indeed, that it may in time become limited to its native wildernesses and to zoölogical gardens.

From Africa we have derived one rather uncommon tenant of our barnyards and fields, the guinea-hen. This creature, though of convenient size, hardy, and commendable from the number of eggs it lays, has never won a large place in the esteem of our rural people, and is now not much kept, except in some parts of the Southern States of this country. The difficulty with this creature, as with the peacock, is that it is not truly domesticated; though it will not betake itself altogether to the woods, it prefers to maintain a half-wild habit. It will not, if it can possibly avoid it, lay its eggs in any place where they are likely to be found by man. Moreover, their rude and little-modulated cries are in the summer season almost incessant, and the din which a considerable flock can produce is exceedingly vexatious. They thus do not fit the [165] needs or comfort of man to the degree which is likely to give them a permanent place among his associates.

The Domesticated Turkey

The last considerable addition to our barnyards has come to us in the form of the turkey. This species has the peculiar distinction of being the only animal form of definite use to man over a wide field which has been contributed from the life of the New World. Although the creature was much hunted by our North American Indians, and is of a type which lends itself to domestication, it does not appear to have become a companion of man until it was taken from the West India Islands to Europe shortly after the discovery of this country. Thence the domesticated form appears to have been returned to this country, where it has been a favorite in a measure unknown in the Old World. Ornithologists deem the Cuban turkey, whence our tame form came, to be specifically distinct from those which are found on the [166] mainland of this continent. Although these kinds are distinguishable by plumage, they are probably only varieties of a common species. This is indicated by the fact that our tame flocks readily intermingle with their wild kindred.

The ease with which the turkey becomes domesticated is remarkable. In this regard the creature may be compared to our cocks and hens. In both cases the tamableness is doubtless to be explained by the fact that the primitive forms dwelt in permanent association, the movements of which were in a way controlled by the adult males, and by the fact that the forms had abandoned the use of wings for wide-ranging flight. The change which has been brought about in the turkeys with their adoption into the human association has been slight. No distinct varieties of breeds have been originated, though here and there the observer may note slight local variations in the coloration of the plumage, which are probably due to varying admixtures with the wild forms of our forests. Thus in Kentucky and other parts of the South, where the opportunities for the intermingling of blood of the tame and wild forms are frequent, the domesticated creatures often resemble so nearly the wilderness forms that even the wary hunter may make mistakes as to whether the bird he sights be fair game or not. Unless carefully watched, a drove of these creatures on the border of the wilderness is apt gradually to return to the wild state, the three or four centuries of life about the home of man not having been sufficient to do away with their ancient love of freedom.

Among the English folk of North America the turkeys found a large place as an element of the food-supply. It has become curiously associated with the Puritan festival of Thanksgiving, an institution which has spread throughout [167] the United States and which has in a way taken the place of the harvest-home festivities of the Old World and bygone ages. It is probable that the relation of this bird to our national festivities has done much to keep it in use in this country. It is a well-recognized fact that it is costly to keep and that the eggs are not desirable for culinary use. The species requires a wide range. It does not do well in the confined conditions in which cocks and hens can readily be maintained. It therefore is not likely to be kept in any region where the agriculture is of a high grade. It is best suited to farms where there are considerable areas of half-wild pastures.

Although the turkey is a truly gregarious form, its mental endowments are of a lower grade than those of most social birds. Their calls are few in number and have little of that conversational quality which we note in those of our ordinary barnyard fowls. Although the males contest the field with each other by personal combats, they are not very valiant, the creatures trusting for favor with the females rather to the parade of their plumage and the pomp of their carriage than to the wager of battle. In the matter of show they are, however, very effective, being surpassed only by the peacock in the splendor of their attire. In their domesticated state they lose much of the beauty which they have in the wilderness, as they do their pristine dimensions. Those who have hunted our wild species are likely to remember scenes where in some forest glade they have beheld a gobbler displaying his graces to an admiring harem. As he struts about with his tail feathers erect and his neck arched back, now and then pausing to utter an exultant gobble, the spectacle is one of the most amusing displays of animal pride which the naturalist has a chance to behold.

The Largest of all Poultry — The Ostrich

[168] Recent experiments in ostrich farming seem to indicate that we are on the eve of introducing into our "happy family" the noblest remaining member of that group of great birds which characterized the life of the later geological periods. As yet the efforts in taming ostriches are too new for us to tell just what the effect of man's skill on the development of this creature will be. It is evident, however,

that the creature can be won from its wilderness state, at least to something like the imperfect companionship with man which has been attained by the guinea-fowls and turkeys. All we know of the variations in plumage of birds indicates that the breeder's art may bring about great changes in the highly decorative feathers for which this bird is to be reared. It is also probable that with the better food which domestic conditions imply, this wanderer of the desert may be brought to attain a very much greater size than it wins in the hard life of its native land. If the form should prove as plastic as that of our ordinary barnyard species, we may indeed [169] succeed in developing a variety approaching in dimensions the gigantic moa of New Zealand, or the æpyornis of Madagascar, those magnificent creatures of the past which passed away just before their native lands were known to our race. The variations in size of the wild ostrich appear to indicate that this interesting result may be attainable.

Next after the cocks and hens the most important birds of economic value have come from the water fowl. In this field there are great opportunities for domestication, only a few of which have been adequately used. The aquatic birds, save for the fact that they are in all cases inspired with a more or less strong migratory humor, lend themselves to the shaping hand of man more readily than most other forms. These creatures have the habit of association in a much more perfect way than our ground birds. They normally dwelt in rather close order and in relations which are necessarily very sympathetic. Whoever has watched the flight of wild geese must have remarked the beautiful way in which they arrange at once for close companionship and for safety in the violent movements which impel their heavy bodies at high speed through the air. In the order of their flight the alignment is more perfect than in the march of trained soldiers. Each bird keeps as near to his neighbor as possible; but manages always to preserve the interval which will insure against a collision of the strong and swift-moving wings, an accident which might well disable them for flight. I have repeatedly undertaken to confound their motion by firing a rifle bullet at the head of the moving wedge. Although the sound of the projectile, if well directed, will disturb their processional order, it never brings confusion. The startled birds sink down or rise above the plane of the air in which [170] their comrades are moving, but they never strike against them.

An Eider Colony

The admirable sense of interval which the wild birds exhibit in their flight is to be seen also when they move over the surface of the water, where the fleet of living forms is always so arranged that each individual does not interfere with its neighbor. I recall with much pleasure an occasion when, from a ship becalmed in a thick fog off the southern shore of Labrador, within sound of the breakers, I undertook to find something about the lay of the land and the chance of harborage by paddling in a small boat toward the shore. I had hardly lost sight of the ship when my boat glided into an assemblage of eider ducks, where the mothers, with their [171] fledgling young, were lazily swimming to and fro, as if to practise the ducklings in the art of swimming. Each brood appeared to have its own space of water, and between each of the chicks there was likewise a less but equally well measured interval. The same features of orderly association, which I have just noted in the swimming and flying of these wild birds, may be seen in a somewhat degraded state in our domesticated varieties of the group. They all indicate in these forms a keen sense of their neighbors and a habit of association based upon sympathetic emotions.

Terns Aiding a Wounded Comrade

The sympathetic quality of our water fowl, at least in that part of the emotion which leads them to be concerned with the afflictions of their species, appears to be more distinct than in the case of our ordinary barnyard fowl. Geese, as is well known, will make common cause against an intruder from whom harm to the flock may be expected. Their [172] simultaneous din when anything occurs to arouse their enmity is commemorated in the ancient myth concerning the aid which they gave in the defence of the walls of Rome. There are anecdotes apparently well attested where water fowl have borne away a wounded comrade which had fallen before the huntsman's fowling-piece. In Smiles's "Life of Edwards" there is an often-quoted story which appears to be trustworthy and sufficiently illustrates this point. A hunter, having shot one of a flock of terns, which fell wounded into the water near the shore, waded in to seize it. Suddenly two of the terns came to their wounded companion, seized him by either wing, and bore him toward the open sea. When these two helpers were weary, the sufferer was lowered into the water, and, in turn, seized by two other birds which were fresh for the labor. Working in succession, these birds carried their companion to a rock some distance from the shore. When the hunter endeavored to approach the rock, yet others of the species seized the cripple and bore him far beyond reach.

Although too much value must not be given to the numerous anecdotes concerning the sagacity of water fowl, the great mass of these stories, as compared with the poverty of the anecdotes concerning the better-known barnyard creatures, seems to establish the fact that their intelligence is much greater than that of the land birds. This superiority can probably be attributed to the fact that their life requires much more definite adaptation of means to ends than in the simpler conditions which are met by the forms which dwell in the fields. The circumstances of their life are something like those of the seals among mammals. They have to do with the conditions of the air, the land, and the water; and as they [173] generally undertake long migrations, the range of the things they have to accommodate themselves to is great, and the effect of their labor is decidedly educative.

Wood Duck   China Goose   Australian Swan   Canada Goose
Some Recent Additions to the Poultry Yard

As yet, from the great number of species of water fowl man has really domesticated but two characteristic groups, the species of geese and of ducks. Swans have been brought to a state where they tolerate the presence of man, though they rarely establish any really intimate relations with him. Some other species, as, for instance, the grebe, have been taught to dwell about the homes of man, accepting food from his hands. It is likely that more of these water fowl would have come into human associations were it not for the fact that they are naturally migratory, and when, after a season of domestication, they join a passing flock, they never return to the place where they have been kept.

Swans

The swan, like the peacock, has been bred for ornament rather than for use. In fact, the bird has no other merit [174] than its exceeding grace. We cannot believe that much pains was ever taken with this creature to break up the migratory instincts which are common in the wild kindred species. We have to suppose that the bird in its pristine form was without the impulse to undertake distant journeys in the winter season, or that it abandoned ancient habits with no great difficulty. We obtain some light on this point by noting the fact that among the migratory species it not infrequently happens that, while the greater number of individuals undertake the annual journey, certain of them will remain on the ground where they were born. Those which remain would be more likely to mate with those which were like-minded than with others that journeyed afar. In this way small local breeds might well be originated which would differ from their migratory kindred not only in the measure of the wandering instincts, but in the capacity for flight which their kindred preserve. There is some reason to believe that [175] this process of selection naturally and somewhat frequently takes place. In certain cases it may lay the foundation of new species, or at least of distinct varieties; more commonly, however, the individuals which have abandoned the migratory life are likely to perish from the severity of climate or the other unfavorable conditions that their mates avoid by their wanderings.

The Original Wild Rock Dove (*Columba livia*) and Some of its Domestic Descendants

Although many of the free-flying birds of the land are or have been kept captive because of the pleasure which men have found from their songs, their grace, or their quaint ways, only one of these has really been gained to domestication. In the pigeon, man has made what is on many accounts the most remarkable of all his conquests over the wild nature about him. While the breeder's art has led many forms, some of them on several divergent lines, far away from their primitive estate, in no other field has it accomplished such [176] surprising results as with the doves. The original wild form of this group is a native of Europe and Asia, where the species

Columba livia, or rock pigeon, is still common, and whence it may be readily won anew to domestication. It is a small, plain-colored, rather invariable and inconspicuous bird about the size of our American dove. In its wild state it dwells in small flocks, nesting by preference in the crannies of the cliffs, and exhibiting no striking qualities which make it seem a desirable subject for domestication. We note, however, that even in this primitive condition the creature has certain physical and mental qualities which have been the basis of its adoption by man as well as of the wide changes which it has undergone at his hands.

It is a characteristic of all the doves that their young are born in a very immature state, and for some time after they come from the egg they have to be supplied with food which has been partly digested in the crop or upper part of the stomach of the parent. For the proper rearing of the brood there is required the assiduous care of both parents. Therefore quite naturally we find among these birds that the pairing habit is well developed, and as they rear several broods each season, that the mating is for life. Although there are numbers of birds in various orders which are accustomed to the monogamic habit, it happens that the pigeon is the only animal which man has ever won to true domestication in which the sexes can be thus permanently united. In the dovecote, however many birds it may contain, the breeder can be always sure as to the parentage of the young which he is rearing. This affords an admirable basis for the practice of his art, which is still further favored by the fact that pigeons reproduce rapidly and the progeny are ready to mate in a few [177] months after they come into the world. Thus the species affords really ideal conditions for that process of selection on which the improvement of all domesticated animals intimately depends.

Turtle Doves

Selective breeding of pigeons began in India, as the records seem to show, more than two thousand years ago. Though other animals have been brought to domestication at much earlier times, this appears to have been the first of them to be subjected to deliberate efforts on the part of their masters, which were intended to bring about in a methodical way certain changes in their forms and habits. The most curious part of this great endeavor which has been applied to breeding pigeons is found in the fact that the ends sought have no utility, but afford satisfaction from the point of view of pure diversion or the gratification of taste. We are well accustomed to the action of such motives upon our flowering plants of the garden, but the pigeon is the only animal where fancy has labored for thousands of years for its gratification. [178] The breeders of pigeons from remote antiquity to the present day appear to have had no definite purpose in all their pains. They have taken the chance variations in form and habit and endeavored to extend these sports of nature by a careful system of mating those in which the singular features were most evident. Thus the fan-tail breed has been developed until the creatures display their unornamental tail feathers with all the dignity with which a peacock shows his marvellous decorations. The pouters have in some unaccountable way learned to take air into their crop; and the habit has been developed by selection until the bird destroys all trace of his original shapeliness, though he seems to take pride in his diseased appearance. The tumbler, probably derived from some ancestor afflicted with a disease of an epileptic character, manages to go through his convulsions in the air without serious consequences and apparently with some pleasure to himself. There are over one hundred less conspicuous varieties, of which only one deserves notice, and [179] this for the reason that it has some possible utility to man and is now much attended to. This is known as the carrier pigeon.

The Giant Crowned Pigeon of India

In early time, before the invention of the railway and telegraph, some ingenious breeder of pigeons, observing the constant way in which these creatures returned to the place where they were bred, invented the plan of using them to convey information. This service was found convenient not only for ordinary correspondence, but was exceedingly valuable where a place was beleaguered by an enemy. In such cases carrier pigeons could often be used to convey information across the otherwise impassable lines. Even in modern times, as, for instance, during the last siege of Paris, these swift and sure flying birds proved of great use in keeping up communications between the people of the invested town and the French armies in the field. Letters in cipher, sometimes photographed down until the characters were microscopically fine, were made into packages of small weight in order not to impede the flight of the bird, carefully affixed to its body, and thus sent away. Very generally these curious shipments came to the hands of those for whom they were destined. The birds can be trusted to fly at night; they retain for a long time the memory of their home, and spare no pains to return to it.

The homing power of the carrier pigeon appears to be a special development of a natural capacity, as is also its swiftness and endurance in flight. Our other breeds and the wild species whence they have all come are not disposed to undertake long journeys; they rarely, indeed, wander far from their abiding places. Our experience with the carriers shows how readily the creatures may be educated to perform feats which [180] they were not accustomed to do in their wild state. Something of the same elasticity of constitution may be observed in the bodies of our pigeons as they have been affected by selection. Not only has the plumage been greatly altered by the breeder's art and in pursuance of his plans, but the form and proportions of the bones have coincidently and unintentionally been greatly changed. So considerable are these alterations that if these creatures were submitted for dissection to a naturalist who knew nothing of the history of the bird, he would have no hesitation in classing them as belonging not only in different species, but as members of diverse genera.

It must be regarded as unfortunate that the experiments which have been made on pigeons have been limited to their features of form, color, and slight peculiarities in their habits. If the breeders had sought to modify the intellectual parts with anything like the insistence which they have given to the development of these bodily peculiarities, we might now have a most valuable store of knowledge as to the limitations of animal minds. The facts gained in the breeding of the carriers show clearly that certain of the instincts of these birds can be readily modified. There is every reason to suppose that their mental capacities in other directions have something of the same pliability.

The English Pheasant

Although the pigeon is the only free-flying form which has been won to intimate relations with man, there are numerous other species of these volant creatures which have been reduced to partial domestication, though they cannot be trusted to abide with us without being more or less completely caged. Experience has shown that by far the greater part of the arboreal birds may be kept and will breed in captivity. [181] From the host of these feathered creatures men have from time to time selected species which grace their habitations by their beauty, their song, or by the sympathetic relations which they form with their captors. Our successes in these efforts toward domestication of these birds have been most eminent with those varieties which in their wilderness state have a well-developed social life, which abide in families or flocks, and have the pairing habit well affirmed. The reason for this has been already indicated. It is due to the sympathetic motive which is developed in such communal life, and is manifested in the friendly relations with each other which the creatures maintain. A good instance of this is to be [182] found in the crows and their kindred, a group of extremely sociable creatures, which are endlessly engaged in chattering communications with each other. All these forms are highly domesticable, and if for any reason they had proved permanently attractive to men they would doubtless have been brought into the state of willing captives.

Although some of the free-flying or tree birds have been kept for their beauty alone, the greater part of them have commended themselves to man because of their voices. It is hardly necessary to tell the reader that the birds, of all animals, are most provided with means of expression through the voice. There is hardly a species which has not a greater range of notes or calls than the most vocal of our wild mammals, and many varieties are impelled to tuneful expression in a measure which no other creature, not even man, exhibits. In most cases these utterances are pleasing to the human ear, for they have the quality which we term musical. Therefore it is not surprising that the most of our captive birds have been chosen for their song.

It seems clear that the song of birds, like their calls—the two shade indefinitely into each other—expresses a sympathetic emotional consciousness of the actions going on about them, particularly of the life of their kind. In general these utterances are directed toward their kindred of their own species. In many cases, however, as among the imitative birds, the sounds which they utter indicate a curiously keen interest in the actions of their masters or other human affairs. The mocking-birds and some other species will, with great assiduity, endeavor to copy any sound which they happen to hear. I well remember watching a mocking-bird which was listening with rapt attention to the noise produced by a man [183] sharpening a saw with a file. The poor bird would hearken with great attention until he thought he had caught the note, and then endeavor to reproduce it. As may be imagined, the measure of his success was small. He was fully conscious of his failure, and would beat himself about the cage in evident chagrin, returning again and again to try the hopeless task.

Wherever the vocal organs of caged birds permit them to imitate human speech they are apt to devote a large part of their labor to this task, paying little attention to other less meaningful sounds. It appears to me that they perceive in a way the sympathetic character of language and therefore take a peculiar pleasure in copying it. It is hardly to be believed that they ever get a sense of the connotative value of words, but it is not to be doubted that they sometimes attain to a certain appreciation of the denotation of simpler phrases. In this task they do not exhibit as much sagacity as the dog, a creature which learns to understand the purport of rather complicated sentences. Nevertheless, their capacity for imitating speech is a fascinating peculiarity, one which has greatly endeared them to bird fanciers.

Those who have observed the talking birds have doubtless noted the fact that their capacity for remembering and uttering words varies greatly. I am inclined to think that in the same species some individuals can do such tasks several times as easily as others. If these speaking forms could be brought to breed in captivity, and something like the selective care were given to their development that has been devoted to the varieties of pigeons, we might well expect to attain very remarkable results. If anywhere in the animal world there is a chance to open communication by means of speech with the lower creatures, it should be here.

The Falconer's Favorite — Peregrine Falcon
[184]

At one time among our ancestors it was accustomed to make much use of the larger hawks in hunting. Curiously enough this amusement, more refined and elaborated than any other form of the chase, has gradually fallen into disuse among Europeans. So far as I have been able to learn, the only region in which it is well preserved is in northern Africa, a country in which the custom was probably introduced from Spain during the occupancy of that peninsula by the Moors. From the literature of this art of hawking, even after we allow much for the exaggeration of unobservant men, it seems certain that the training of these fierce birds was carried to a point of singular perfection. The creatures learned to do their duty in a very skilful way, and they readily acquired habits of obedience, under circumstances of excitement, more perfect than those which we succeed in instilling in any animal but the dog. When we consider the natural [185] qualities of the hawk, and note that when well trained he flew at only the designated game, and came back to the master when a bit of hide or other lure was thrown into the air as a signal, we may fairly believe that the creature displayed an extraordinary fitness for receiving instruction. The facts are the more remarkable because these hawks were not bred in cages, but were taken from the wild nests; so that there was none of that gradual accumulation of inheritances under the conditions of selection which have brought about the obedience of our really domesticated animals.

The remarkable way in which the art of hawking has disappeared from our civilization deserves more than a passing notice, though it appears to be inexplicable. It is evident that it was a tolerably ingrained habit, at least among the English-speaking people, for it has left a very deep impress upon the language. There are far more phrases derived from the custom than can be traced to any other of the sportsman's arts. At least one of these collocations of words which has escaped from the minds of grown people still holds a place among the boys of this country. When two lads are fighting we often hear the bystanders say, by the way of encouragement to one of the contestants, "Give him jesse." The use of this curious phrase prevails in all parts of the United States, but after much inquiry I have failed to find a trace of it preserved in England. There seems to be little doubt that these words are due to a custom of beating a hawk which failed to do its duty with the thongs or jesses by which it was attached to the wrist of the falconer. Giving another jesse thus came to be equivalent to giving a person a strapping.

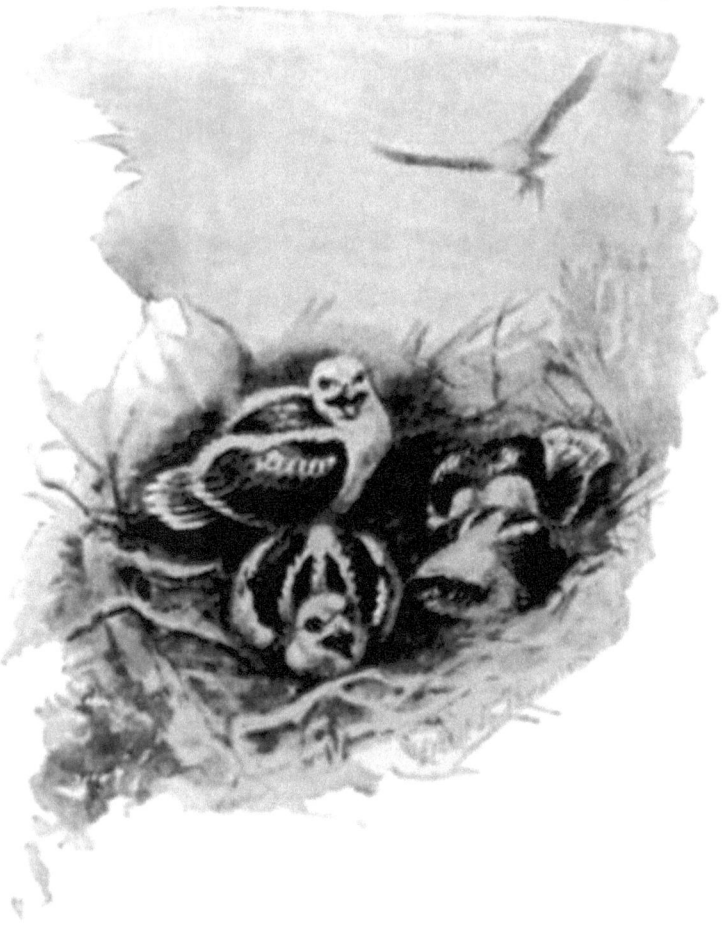

The Bandit's Brood

Whatever may have been the reason for abandoning this beautiful and in a way noble sport, its disuse must be deemed [186] most unfortunate by all the students of animal intelligence, for it has deprived us of precious opportunities in the way of observations on the mental peculiarities which exist in a most interesting group of birds. In these days, when there is a fancy for reviving the customs of our forefathers, it might be well for some persons of leisure to give their attention to restoring the arts of falconry. Enough of the practice and of the traditions is left to make it an easy task to reinstitute all the important parts of the custom. Moreover, those who essayed the matter would have access to a much greater range of rapacious birds than our forefathers, who had to content themselves with the limited number of wild species which inhabit the continent of Europe. Especially on our Western plains, where game-birds abound and the country lies wide open, sportsmen would find an admirable field in which to follow the bird they flew. Not only would the restoration of hawking give us a sport much more skilful and refined than the fox chase, but it would reintroduce the cultivation of the only creature which, having once been [187] brought to the service of man, has been permitted to return to its ancestral wild life.

The most striking and by far the most interesting quality exhibited by our birds is found in their sympathetic motive. In this spiritual quality, so far as it relates to their own kind, the feathered creatures are clearly in advance of all other species, including even man. A single fact, one of great generality, will serve to make this statement clear. Among the birds we find the only cases of true marriage which are known in the animal kingdom. In the greater number of the species the union is for a season, but among many it is for life. In the case of certain varieties of paroquets, the union is so indissoluble that, according to common report, a report which seems much better verified than the most of those concerning the habits of animals, neither member of the pair will survive the death of the other. Man, with all his striving towards a better social state, has, as a whole, not yet attained to the enduring affection for the mate which is evinced by the greater part of the birds.

In this same connection, we may note that the æsthetic appreciation among the birds appears to have attained a far higher level than it has won in any other creatures. There can be little doubt that the exquisitely beautiful plumage, the unparalleled shapeliness of form and grace of carriage, as well as the melodies which are uttered by so many species, all owe their development to a process of sexual selection which has led the discerning females to prefer the more ornamental of the males who sought them as partners. If any one will examine the exquisite shapes and gradations of color which are exhibited in the tail of the peacock, or of the lyre-bird, or even the coloration of the game-cock, he may perhaps imagine [188] how prodigious must be the development of the æsthetic sense in these species, in order that it may take account of every little betterment which leads towards more perfect beauty. As it will take the generations of æsthetes many generations before they are able to "live up to" the level of their culture which is attained by the peacock's tail, it is not unreasonable for us to hold that in the appreciation of simple beauty in form and in color, the birds are far ahead of ourselves. It must not be supposed that our æsthetic culture is to be reckoned below that of birds, though in our case the work embodies the delineation of ideas, while in the birds it is a matter of pure ornament. Nevertheless, taking the evidence which shows the way in which these creatures appreciate beauty in the three realms of form, color, and sound, it seems to me clear that while their intellectual life is low, their purely emotional experiences are probably more vivid than those of ordinary men.

As the joy of life is, in the main, even in ourselves the result of emotional experiences, we may fairly reckon, even on *a priori* ground, that the birds win a measure of happiness, though it be that of an unconscious kind, which is granted to no other living beings. Psychologically described, they might well be termed the group built for joy. Their bodies are, on the whole, the best constructed of all animals, except the insects. They suffer little from disease. We all see that their intercourse with each other is freer and merrier than that of other creatures. The wide range of their notes shows that in most forms they appreciate every little difference in the pleasure-giving changes of the day or the weather. They rejoice in the coming of each morning; they are sorrowful with the advent of each evening. They echo [189] the distress of their kind in a readier way than any other forms. He is indeed a poor naturalist who overlooks this trait; for however deeply he may have delved, he has not won the jewel unless he appreciates this element of an unending joy which the bird-life continually offers him. From that life we may well believe that man is hereafter to derive some great and fruitful lessons.

[190]

USEFUL INSECTS

Relations of Man to Insect World.—But Few Species Useful to Man.—Little Trace of Domestication.—Honey-bees: their Origin; Reasons for no Selective Work; Habits of the Species.—Silkworms: Singular Importance to Man; Intelligence of Species.—Cochineal Insect.—Spanish Flies.—Future of Man relative to Useful Insects.

Although the relations of man to the insect world are prevailingly those of hostility, there are a few of these multitudinous creatures which have been more or less completely adopted into his great society. Although not more than half a dozen out of the million or more species in this subkingdom have thus been brought to the uses of civilization, the forms are interesting not only for what they give, but for the promise of further contributions when this great problem of winning help from the insect world receives adequate consideration.

As a whole, the insects are not well fitted to serve the needs of man. Owing to certain peculiarities in their organic laws they, fortunately for ourselves, are very limited in size. Although some of them afford savory food and are occasionally eaten by savages, and even by civilized folk when pressed by hunger owing to the famines which the invasions of these animals occasionally produce, they can never be of any value as sources of provisions, except through the stores which they accumulate in the manner of the bees. All that we have won, or are likely to win, from this realm is from the filaments which the creatures spin, the wax or honey which they accumulate, the coloring or [191] other matters which their bodies afford, or the help which they may give us in our struggle with invading species of their class.

Probably the first insect to be brought into friendly relations with man was the honey-bee. This creature, like the most of our domesticated animals, is a native of the great continent of the Old World, though it has now been conveyed to all the flowery lands of the world where the season is long enough for it to win its harvest. In its wild as well as in its tame state the honey-bee dwells in one of the most perfect and highly elaborated of insect societies. It is a member of the group of membranous-winged insects known to naturalists as *Hymenoptera*, an order which includes all the elaborate societies of the class except the colonies of white ants. It is characteristic of all these colonial insects that, from the experience of ages, they have learned the great principles of the division of labor and of profit sharing towards which mankind are now clumsily stumbling; the great work which their societies are able to do is accomplished by a complete specialization of function and a perfect share in the commonwealth. So far has this elaboration gone, that in the bees the work of reproducing the kind is allotted to forms which do no labor; all the work of the hive being effected by individuals which are sterile, and whose sole function it is to toil unendingly for the profit of the great household.

While the greater part of the kindred of the bees either construct the nests for their young in the manner of our wasps or hornets, building them entirely in the open air, or excavate underground chambers in the fashion of our bumble-bees, our domesticated form at some time in the [192] remote past adopted the plan of choosing for its dwelling-place some chamber in the rocks, or cavity in a hollow tree which could be shaped to the needs of a habitation. Owing to the size of these cavities, they were enabled to form societies composed of many thousands of individuals; while the species which adopted nests, in other conditions, were much more limited as regards their numbers. Thus the bumble-bee, which abides underground, dwells in very small communities, probably for the reason that the conditions of the soil it inhabits make it difficult to excavate and maintain large rooms. It is this habit of resorting to hollow spaces, as well as the instinct to store up honey in wax cases, which has made the common bee valuable to man.

[193-4]

Feeding Silkworms with Mulberry Leaves in Japan

At best the opportunities which the wilderness affords, in the way of fit dwelling-places for the swarm which goes forth from a hive, are much less than can readily be provided by art. In almost all cases the wild bees have to expend a great deal of labor in searching for a fit residence; and after such is found it requires a great deal of toil and expenditure of the costly wax in order to shape the cavity so that it may comfortably accommodate the multitude, and be reasonably safe from the attacks of other insects. Thus it has come about that the bee has, in a way, welcomed the interference of man with his ancestral conditions; and, though the species exists in the wildernesses of its native land, the domesticated varieties have so far taken up with man that in other countries they do not wander far from the limits of civilization. Now and then an uncared-for swarm which cannot find accommodations about the parent hive will betake itself to the wilderness; though it generally continues to seek [195] sustenance from the abundant flowers of the tilled fields where it finds species, such as clover and buckwheat, from which it has been long accustomed to win the harvest of pollen and honey.

In North America the honey-bees, which were brought by the early settlers, and which had been kept on the frontier by the pioneers of our civilization, have always extended, in wild swarms, a little distance into the wilderness. But, at most, they appear to have wandered only for a few miles beyond the homestead, going no further away than would permit their use of the cultivated plants. The aborigines early learned to regard the insect as the *avant courier* of European men. When they came upon an individual of the species they always knew that some white man's dwelling stood nearby. Those who are familiar with the solitudes of our Appalachian forests must often have remarked, in the stillness of a summer day, the hum of a swarm from some forest or domestic hive in its search for a dwelling-place. Those who have followed up the movements of these migrating colonies have had a chance to perceive how long is the search before they find a fit abiding place. Doubtless by far the greater part of these searchers for a home fail of their quest, and the wandering swarms perish without finding a suitable shelter.

In certain kinds of woods, as, for instance, those occupied by pine trees or other species which do not develop spacious hollows in their trunks, and where there are no crannied rocks—all the swarms which seek habitations there are foredoomed to destruction. If by chance the colonies wander too far, they generally find the wilderness so ill provided with plants which may furnish them with the sources of wax, [196] honey, or other necessaries, that they cannot maintain their life. Thus it is that the bee, though domiciled with us rather than domesticated, has become united in its fortunes with civilization. In this position they have shown a remarkable adaptation to extremely varied conditions. They can withstand any climate which permits the development of the vegetation to which they need have access, provided the growing season continues long enough to accumulate their store. In the tropical lands they harvest so little honey that they are not profitable to man, and in the high north they need all their summer's accumulation to maintain them through the long winter. Thus, though they may range almost as far as man through the gamut of climates, they are profitable to their masters only in the middle latitudes. They commonly do not do well close to the sea, and cannot be kept on inconsiderable islands for the reason that they are, in their wanderings, likely to be lost in the waters.

The bee, like the other social insects, evinces a wide range of instincts which are intimately related to the economy of the hive; but these motives appear to be of an unchangeable character. They show no tendency to undergo the modifications which we observe to take place in our birds and mammals when they are brought under the influence of man. The only case in which they show any distinct effect from their contact with man is found in their evident recognition of those who care for them. They soon learn that their master is not to be feared, and, therefore, need not be resisted; but, beyond this dumb acceptance of a situation, they exhibit no trace of sympathetic recognition of our kind. It is clear that their mental endowments, though considerable, are very much more remote from our own than are those of the vertebrated [197] animals with which we have formed a friendly association. Moreover, the type of life of the creatures in a way excludes them from any kind of share in human society. Each of them is, from its birth to its death, entirely devoted to the interests of its little commonwealth. Every impulse of their being relates to the economy of their hive. While we know little about instinct, we know enough of its manifestations to state that the real unit of this species is not the individual insect, but the colony to which it belongs. The separate form is hardly more than a bit of machinery so arranged that it may operate at a distance from the engine of which it forms a part. On this account it appears to be impossible for us ever to attain to any kind of sympathetic relations with these creatures.

Even more important than the bees are those insects which, in their immature state, yield us silk. The so-called silkworms, like the bees, originated in Asia, and have long been in the care of man. Beginning their experiments in spinning with the wool of animals and the various accessible vegetable fibres, men have ever been seeking materials which could serve them in the weaver's art. At one time or another they have tried an exceeding variety of materials; in modern days more than a score of insects have been experimented with in the endeavor to obtain fibres which could be turned to use. So far, however, the *Bombyx mori*—the form which, as its specific name indicates, feeds upon the leaves of the mulberry tree—is the only one which proves really serviceable. The advantages of this species are found in a peculiar assemblage of qualities, each of which is necessary to make it fit for the ends it attains at the hand of man.

[198]

The mulberry silkworm can readily be bred in confinement. The eggs are easily gathered and preserved, and are so readily kept that they may be sent the world about. At a given temperature they with infrequent failures hatch; and if sufficiently fed with the fresh leaves of the mulberry, will in a short time attain to as perfect a development as though they grew, not in close rooms, but in the open conditions of the trees. When of adult size, the grubs proceed to spin themselves in, forming a thick cocoon composed of threads of a material which, though as soft as paste when emitted from the body, hardens so as to form a strong and even thread. If the insect be allowed to remain for a sufficient time in the cradle which it has spun for its second birth, the body within the chrysalis case will proceed in a manner to dissolve; and in the milky fluid thus produced, where only faint traces of its former state remain, the beautiful image or perfect form will arise. In the economic use of the creature, however, except as far as a supply of eggs may be desired, it is necessary to prevent the completion of its development; for in escaping from the chrysalis case, the butterfly cuts many of the delicate threads, so that the silk is made unserviceable. It is necessary to wind it off before the insect escapes. In this part of the work we notice the most perfect adaptation of the creature to the needs of man. While the silk threads from the cocoons of other species which might prove of value cannot be easily reeled off, those of the silkworm, when placed in hot water, readily separate, and can be gathered in a condition for spinning. Thus, while some success has been attained by carding the cocoons of other species, thereby making a fibre which has a certain utility, the silkworm alone yields material fitted for delicate fabrics.

[199-200]

The Farmer's Apiary

[201]

At the present time in Europe, Asia, and America there are probably not far from ten million people who depend in large measure upon the product of the silkworm for their livelihood. Although the product of their industry and that of the insects combined is not nearly as indispensable to man as those which are won from the hair of animals or the fibres of plants—for silk is a luxury rather than a necessity—the value of the work done by these humble creatures is greater than that effected by the largest of our domesticated animals, the elephant. If the philanthropic economist were forced to choose which of these creatures should pass from the earth, he would have to accept the loss of the greater and far nobler animal.

So far as regards their intelligence, the silkworms are much below the level of the bees. Though they dwell in an aggregate way they have scarcely a semblance of social order, and are without the wide range of peculiar instincts which we invariably find among the commonwealth animals. The order of *Lepidoptera*, in which these creatures belong, though the most beautiful, appears to be from an intellectual point of view the least advanced of our insects. Their instincts are all on a low plane; they have no kind of mutual labor, and however much advance we may make by selection in developing their bodies, there is no reason to expect that we shall affect their intelligences.

The cochineal insect, a species which has the habit of feeding upon the cactus, is used for a dye stuff, for which service the brightly colored body is appropriated. Although the creature is deliberately planted where it is to feed, and thus is in a way submitted to culture, it cannot fairly be said to have been entered in the domesticated circle of man. In [202] a similar way the so-called Spanish fly—which really belongs among the beetles—whose ground-up bodies are used for producing blisters, is merely appropriated to our use without any process of subjugation. The fact remains that, so far as our dealings with the insect world have gone, we have really won but two of the million or more of forms to captivity; and our relations with these have nothing of the humanized nature which marks our intercourse with truly domesticated creatures.

Small as are the lessons which we may read from our experience with the honey-bee and the silkworm, they appear clearly to indicate that, while we may expect to do little with the intelligences of insects, we may fairly reckon on a great field for accomplishment in the way of changes in their bodily constitution. In the case of the bees the facts show us that in particular conditions of climate or other surroundings a certain amount of variation takes place, and by proper selection either of queens or swarms it may be possible considerably to extend the value of these animals. The task is beset with difficulties for the reason that, while in ordinary selective breeding we deal with individuals, we have, as before remarked, in this species to regard the hive or colony as the unit and to make our selection with reference to the qualities of that colony as a whole. Nevertheless, with the constant advances in the skill of our economic selectionists, there is reason to expect that our bees may be progressively improved. On the other hand, there is the chance that the progress of chemical discovery may enable us at any time to manufacture honey in the artificial way and of a quality indistinguishable from that produced by domesticated bees; in which case these [203] captives, at best troublesome, though most interesting, will probably disappear from the human association.

With the silkworms, variations can be more readily brought about; for, as is the case with other animals, the individuals can be paired. The efforts at selection already made show that valuable characters can be thus accumulated, though not with the success which attends the efforts of a like nature made in the case of our domesticated mammals and birds. In common with other animals—indeed, we may say, with all organic life—the silkworms vary perceptibly in different parts of the world to which they may be taken. Thus, when reared in California it is said that this insect develops more strength than it exhibits in Europe; and the eggs which it lays there produce stronger insects, which in turn yield larger cocoons than the individuals born in Italy or France. With such a basis for the selective art as the variations of this insect afford, there seems no reason why it should not afford a good field for the work of the breeder's art.

[204]

THE RIGHTS OF ANIMALS

Recent Understanding as to the Rights of Animals; Nature of these Rights; their Origin in Sympathy. — Early State of Sympathetic Emotions. — Place of Statutes concerning Animal Rights. — Present and Future of Animal Rights. — Question of Vivisection. — Rights of Domesticated Animals to Proper Care; to Enjoyment. — Ends of the Breeder's Art. — Moral Position of the Hunter. — Probable Development of the Protecting Motive as applied to Animals.

It is well to note the fact that, in considering the rights of the creatures below the level of man, we are dealing with a question which does not seem to have entered into the minds of the ancients. Such old phrases as "the merciful man is merciful to his beast" indicate that cruelty to the domesticated creatures was, in a way, reprobated by the ancients; but not until well on in the present century do we find any indication that reason had come to the help of pity in an effort to frame rules having the weight of law and the support of sanctions, either those of public opinion or the more direct penalties of the courts, to limit the conduct of men towards the lower animals. The great tide of mercy and justice which marks our modern civilization had first to break down the grievous and strongly founded evils of human slavery. Having effected that great work, the sympathetic motives are moving on to a similar conflict with the moral ills which arise from an improper treatment of those slaves of a lower estate, the domesticated animals.

It is impossible to see our position in relation to the matter of the rights of animals without looking somewhat [205] carefully into the intellectual and moral steps which have at length brought us to the consideration of the question. First let us note that while the rights of their fellows have been impressed on men by the precepts of religions, particularly by those of Christianity, the rules of conduct which guide us in our contacts with beings below the level of our species have never been determined by the canons of our faith, for the reason that they are the product of very modern conditions; they are the thought of our own time. New as are these tenets, however, they may fairly be received as but the last though not the final expression of that most interesting of all natural series — the succession in the development of sympathy which, step by step in the progress of organic life, has led from the original dull insensitiveness of the lower animals upwards to the outgoing spirit of man.

In the lower stages of animal life we find no traces of appreciation of the neighbor except those which necessarily relate to the selection and capture of food and perhaps to the selection of mates. Further on in the process of development we note the love of offspring, and, as a consequence of that love, the growth of the family sense, which rarely is maintained beyond the time when the young can shift for themselves. Among the species of the higher groups—certain insects, the greater part of the birds, and the nobler of the mammals—the instinct of the family is extended until it includes the tribe, or perhaps goes yet further and leads to a certain kindliness to all the individuals of the race. Thus it comes about that the individuals of many species below the level of man will respond to the cries of their kindred though they may never have had a chance to know them. There is in these cases a sympathetic bond that [206] binds the kind together. It is with this condition of the sympathies that the task of their further evolution is transferred to man. Inheriting as he does the essential motives of the lower beings through which he came to his present estate, man proceeds to deal with them in a manner which is determined by the peculiar rational power which belongs to him. In place of the blind following of the emotions which characterizes the sympathetic movements of the lower animals, we find that even among the most primitive and lowly savages rules of conduct are instituted which serve to direct the ways in which the individual shall act with regard to his fellows. In almost all cases these rules are much intermingled with the religion of the people; usually they rest upon a body of advancing public opinion which amplifies the motives and, in turn, is enlarged by their growth. As time goes on and the folk attain the stage of records, these rules of conduct become definite laws which at first are based on religious ordinances; but in time they are, in the latest stage of social growth, brought into the state of ordinary statutes which, while they may have some religious sanction, are supported by the machinery of the secular government.

After the first rude work of shaping the body of ancient experience into law was done, there remained the larger and more difficult task of continuing the development of the sympathetic motives with a corresponding amplification of customs and statutes so that the steps of advance should be duly embodied in these rules of conduct. The stages of this purely human attainment have been slowly taken, the onward way has been effectively won but by few peoples. A part of the slowness in advance in the enlargement of [207] the sympathetic motives beyond the stage which has been attained in the life below the human grade is to be accounted for in the fact that no sooner are laws formed than they become in a way sacred. If they be cast in the religious mould their sanctity may be such that they are almost beyond the reach of modification; even when they are secular the reverence for the wisdom of the forefathers naturally leads men to regard them as the ark of safety. Thus it has come about that the codification of the ancient sympathies, won by experience in the pre-human time and in the early life of man, has led to the institution of a barrier which makes further advance a matter of difficulty — one which, in the case of most peoples, binds them firmly to the past, arresting their sympathetic development at a point which it had attained when their laws were framed. This is, indeed, the position of nearly all the peoples except those of our own Aryan race.

When the conditions of a people are fortunately such that they may continue their sympathetic growth, they proceed to carry onward the process of sympathetic enlargement, modifying their laws to suit the gains in understanding which come with this growth. It may be noticed that the development takes place most readily where the rules of conduct are embodied in statute law; for this law, being the evident result of human action, is manifestly alterable in a way that cannot be taken when the prescriptions are supposed to rest on divine commands. Under such conditions of statute law men are freer to advance than they can possibly be where the rules of action are in the form of revered precepts, such as guide the peoples who are accustomed to base their action on the books which they esteem as sacred. Endowed [208] with this element of freedom, the peoples of our own Aryan race—and, fortunately, the most advanced of all its varieties, the English-speaking part of the folk— have, by the divine impulse towards moral advancement, been led to make a great extension of the sympathetic motives. The first step in this direction seems to have been towards the mitigation of the horrors of war, which of old meant the slavery or slaughter of the prisoners. Under the dictates of the developing spirit of mercy and without written law, these brutal actions have been limited until the dogs of war are allowed to rend only in the hour of battle. In this day the man who slays the wounded or robs the dead is esteemed an outlaw. The same beneficent motive was next extended towards human slaves. In this matter English people led; and to them it was almost altogether due that this evil has come nearly to an end except among the Mohammedans, who are bound as in chains to their sacred books and cannot win their way to progress through statutes. In a like manner, in the care of the poor, of prisoners for debt, and even of malefactors, our English folk on both sides of the Atlantic have led in the ongoing towards a higher moral estate.

The last great excursion of sympathy which has characterized the English Aryans—one dating its beginning to this century—is that relating to the rights of our domesticated animals. This has come about, like the other movements, in a way unconsciously. Prophetic spirits have seen beyond the vision of their fellows; they have given their messages, which have found an echo in the souls of men. The motive originated in the recognition of the essential likeness of the minds of the lower animals to our own. But it has been greatly reënforced by the teachings of the naturalists to the [209] effect that all the life of this sphere is akin in its origin and that our subjects are not very far away from our own ancestral line.

It is characteristic of sympathetic movements that, while they are slowly prepared for, their final development is very rapid. Thus it has come about that within one hundred years the conception of the rights of animals has advanced with almost startling rapidity. No other moral gain has been made with such speed or has so rapidly become a part of the property of civilized man. The steps are those which have been taken in all the other great moral advances: at first there were but a few who, in the manner of the skirmishers of armies, set the standards far on in the new ground; gradually the less ardent win their way to them, only to be led the further by their natural guides. As the great advance is still making, it is difficult to see how far it may attain; it is, however, easy to recognize some of the important gains and to foretell the path if not the field of full accomplishment of the conquest. A century ago a man, so far as the law was concerned, owned his living chattels as he did the inanimate things of his property. He could torture or slay them as whim or malice might dictate; there were no limitations by statute, and public opinion, where it might reprobate, was too weak to influence his conduct. Now the statute books of all countries which are moving in the path of moral advance show that public opinion has attained the point where it begins to formulate itself in statutes which restrict the relations of men to their domesticated animals—or, in other words, endow them with definite rights. He may, of course, force them to do him their fit service; he may at his need slay them; but he must exercise his authority without brutality; he must, in form at [210] least, be merciful unto his beasts. With this limitation the rights of domesticated animals began to exist.

At first sight it may seem unreasonable to found the rights of dumb beasts on the embodiment of public opinion in the law, and this for the reasons that many persons have held, that rights have an establishment in the ultimate moral constitution of the world. It may be granted that even before man or even life existed in the universe there were certain logical moral principles which were destined to take shape when the creatures to which they were adapted came to be; but such speculations are fanciful and do not much concern those who are dealing with the problems of the barnyard. We may, to bring the matter nearer, say that the slave of half a century ago had a right to be free; but this right, in all practical senses, meant only that certain people very much disliked to see him enthralled.

So far, by successive stages, first by accumulated public opinion and then by its embodiment in statutes, we have won a measure of protection to subjugated animals which tends to save them from the extremer forms of cruelty. The question now is as to the advances which may be made in the time to come. It is evident that these advances, so far as the domesticated species are concerned, will have to be limited by the needs of man. We cannot ever expect to have the reverence of the Hindoo for the lower animals, for the reason that his state of mind is based on the preposterous supposition that the beast contains the spirit of a man on its way through the cycles towards perfection. We must continue to burthen, tax, and slay; but we may fairly be required to inflict no unnecessary suffering. In this process of amendment we shall undoubtedly before long come to the point [211] where we shall demand that these animals shall be lodged in a wholesome manner and so fed that they may be fit for their tasks. We may, in a word, consider their well being so far as it is consistent with the well being of mankind, and in so doing we shall demand some personal sacrifice from the owner where such is clearly demanded to maintain the principle of the law.

As in all other great sympathetic movements, the leaders of the advance in the matter of the humane treatment of animals are occasionally unreasonable in their demands — it may well be held that the prophet has to be unreasonable in order to attain his goal; hence it has come about that the demands of these admirable people are often beyond the bounds of things that are practicable. Fire-horses, however ill, should be made to do their duty, even if it costs them any amount of suffering; even as the artillerymen should, if the occasion calls for it, rush their teams, though they know that the poor beasts are to die at the goal. In a word, the only and supreme test of our relations to these subjects is the well being of man considered from the higher point of view. This principle we apply to our own kind; we are justified in like action in case of the brutes. In this consideration, the offence to the feelings of man which is caused by any act of cruelty, however necessary, deserves its due weight.

The most serious matter connected with the question of the rights of animals which is now under discussion relates to the use of these creatures in the investigative work of the naturalist, or in the repetition of the processes and results of those inquiries before students. Although all judicious people are likely to welcome the exceeding reprobation with which many philanthropists visit the vivisectionists, and this [212] for the reason that the state of mind shows a rapid advance of the sympathetic motive, they are likely to question the sound foundation of the objections that are raised to experiments with animals, made for the purpose of discovering of displaying the truths of nature.

So far as the work of research into the phenomena of life is concerned, there can be no question as to its importance or as to the fitness of sacrificing the lives of the lowlier creatures in any way that may be necessary for the advancement of knowledge. In the last half century there has been an improvement in the treatment and prevention of diseases so great as almost to defy adequate description. To take only the last of these precious gains, that in relation to the treatment of diphtheria, the gain has been such that although the process is not past its experimental stage the reduction of the mortality in hospitals where the remedy is used has lowered the death rate from above fifty to about fifteen per cent. of the cases. Yet this result rests upon a vast amount of experiment which has cost suffering and life to the lower animals; and to produce the remedy which is used, horses have to be innoculated with the disease, and thereby much pain is inflicted upon them. Weighed as against the life of a human being, a host of the lower creatures must count as nothing. As all human advancement depends upon the dissemination of knowledge, it is difficult to see any objection, from the point of view of justice, to the use of the lower creatures to accomplish this end. The only real point in the matter is as to the effect of such scenes on the minds of young people; yet they have to be accustomed to behold the processes of destruction of life which are everywhere going on about them. The gardener maintains his work by endless [213] slaying. Our tables bear the products of the slaughter-houses. While the anatomist's work may be revolting, it is only so because his tasks are done deliberately and for a purpose that is not yet properly appreciated.

It is a curious fact that many a person who enjoys hunting or fishing, and who slays or maims with much pleasure and to no substantial profit, is horrified to see a student dissecting a living frog, guinea-pig, or cat, in order that he may learn new truths or himself behold what others have discovered. Of the two aims, momentary pleasure or intellectual profit, which is the nobler? In which work is the mind the most likely to become careless as to the rights of the dumb beast? To my understanding, the present turn of sympathetic people against vivisection indicates that the movement of the emotions has, as is often the case, been diverted from the fittest path. So far from natural science tending in any way towards cruelty, it has been the very guide in the development of the modern affection for living beings. By showing something of the marvels of their structure and history, it has increased in a way no other influence has ever done the conception which we form as to their dignity and the wonderful nature of their history. It is in the true interest of mercy to disseminate in every way we can knowledge as to the real nature of animals, leaving this knowledge to bring forth the good fruit which it ever bears. In this connection it should moreover be said that the naturalist, like the surgeon, instinctively seeks to make his work as little painful as may be to the subjects of his experiments. In almost all cases, the animal is made unconscious. Moreover, all we know of the life of the lower animals leads us to suppose that while they suffer much as we do, their pains are of a physical sort, [214] and unassociated to any great extent with the large fears and anticipations which in the case of man form so considerable a part of his torment when in face of death.

The question of vivisection is but a part, indeed a very small part, of the much larger problem as to the relation of men to the lower life which is about them in their fields and in the wilderness. An approximate census of the species now on the earth shows that the number is between two and three million. In the presence of this host, we have to recognize that each of the innumerable individuals in its lifetime is a record of toil and pain the history of which extends backward to the beginnings of life. In this wonderful living world man has trodden ruthlessly, for the reason that he has no sense as to the dignity of the field. In the manner of a vandal, he has slain for profit or sport. He has been so effectual a destroyer that species, genera, and even families of animals have been ruthlessly swept away. The revelation of natural science, of the men of the knife who are so hated by some well-meaning but misdirected people, have now and only in our day brought us to a point where the sense of nature in its organic aspect begins to penetrate the minds of men. The revelation is so vast in its contents and its imports, the conceptions which rest upon it are so greatly enlarging to the human soul, that we may be sure of the wide and swift extension of the new light. It cannot be questioned that the clearer insight will rapidly change the attitude of men toward all living beings. We can in a way discern some of the conceptions as to the rights of the other life which will be enforced on mankind.

It is likely that the first step into the new field of human duty, due to our better understanding as to our place in [215] nature, will be in the direction of a greater care as to our domesticated forms. While we must continue to make their lives subserve our own, we may well insist that they should be properly housed, and have what it may be possible to afford them in the way of their primitive joys, which come from the sun, the air, and their natural food. No one who has seen a long-stabled horse made free of a field can have failed to note the intense pleasure which he takes in returning to something like his natural conditions. Many a cow stable with its foul conditions inflicts more and more enduring torments than all the vivisectionists that some misguided philanthropists are fighting; yet because of the novelty of the naturalist's work these attend to the new scene and neglect the ancient abuse. Among these evils which are to be corrected we may also account that which arises from the unguided development of what are called fancy breeds. Thus among our horned cattle, the Jerseys have been brought to a point where, from the iniquitous inbreeding, which is against what may be called the morality of nature, they are fearfully subjected to tuberculosis. The punishment for this insensate performance comes back upon mankind in the dissemination of consumption; but unhappily it does not visit the people who are responsible for the development of this breed. A like, though less considerable, evil is shown in the fancy breeds of dogs, pigeons, and some other petted animals, where for amusement and as an indication of his power man has raised up many decrepit and sickly varieties, which are not likely to have a fair share in the pleasure of life which their natural breeding insured them.

The observant naturalist of the field has the sense—at least he has it if he be endowed with a little imagination—of [216] the immense pleasure which life gives to most wild animals. That instinctive, and in its foundations utterly irrational and animal joy which men have, or should have, in their day, is part of the birthright of all sentient beings. As yet we have not recognized that this privilege of enjoyment should be confessed. We do not hesitate to slay or maim for mere sport. It is true that some of the ancient forms of this sport, such as bull-baiting and cock-fighting, have been condemned, but the best of men go afield with the gun to slay for pleasure. In a measure they keep up the pretence that they are in some way contributing to the needs of the larder, but so far as needs are concerned the pretence is mostly idle. It seems to me clear that in shaping our sympathetic relations towards animals in the light of our present knowledge, the huntsman will soon become unknown in civilized life. So long as men looked upon animals in the childish, ignorant way, viewing them as utterly commonplace things, hunting or fishing, for the reason that they rested on a foundation of ancient emotions, might well be indulged in. But to the man who knows what science has to teach him, and who discerns the marvels which the animal form enfolds, the destruction of such objects, except for need's sake, is sure to be painful. I judge this from my individual experience. In my youth I was very fond of hunting, and could even wring the necks of wounded birds without trouble of mind. A better sense of what life means, a sense which is no better than that to which all educated men are soon to attain, has made such work very repulsive to me.

When the knowledge of our time is so brought down among the masses of men that it may afford the foundations [217] for appropriate enlargement of the sympathies, the result will doubtless be a great movement towards enlargement in public opinion which credits the lower life with what we term rights. The most important result of this movement will be the creation of a sense of duty by this life. It is said of Mohammedans that they hesitate to tread upon a bit of paper lest it bear the name of God. We know now full well that every living creature in this world bears the stamp of a Providence which has acted from all time, and that we, so far as our own advancement will permit, are morally bound to allow this life to go forward on the appointed way.

[218]

THE PROBLEM OF DOMESTICATION

The Conditions of Domestication; Effects on Society; Share of the Races of Men in the Work. — Evils of Non-Intercourse with Domesticated Animals as in Cities; Remedies. — Scientific Position of Domestication; Future of the Art. — List of Species which may Advantageously be Domesticated. — Peculiar Value of the Birds and Mammals. — Importance of Groups which tenant High Latitudes. — Plan for Wilderness Reservations; Relation to National Parks. — Project for International System of Reservations. — Nature of Organic Provinces; Harm done to them by Civilized Men. — Way in which Reservations would Serve to Maintain Types of the Life of the Earth; how they may be Founded. — Summary and Conclusions.

The advance of mankind from the primitive savagery has been accomplished in many ways. Among the various paths of onward and upward going, however, we trace three which have served greatly to secure the elevation of our estate. First of all, culture came through the use of the hands in the development of the simpler arts. Next, these arts led men to search the stores of the wilderness and of the under earth for materials which could serve them in their advancing crafts. The third important stage in their ongoing was attained when they began to subjugate the animals and plants of the wilds, bringing the creatures to abide in and about the households. Although in general this was the last great step to be taken in the beginnings of civilization, it was on many accounts the most important.

Until men began to domesticate the forms of the wilderness, it was impossible for them to rise above the grade of savages. Their supply of food was necessarily in such a measure limited that their societies had to remain small and [219] they were given to much wandering to and fro over the earth. Moreover, they had only the strength of their own hands for all the work of life. It was not until our kind began to form a society of other species about their homes that the foundations of civilizations were firmly established. The home, indeed, may fairly be said to be the product of the conditions which the process of domestication brought about. As distinguished from the temporary hut of primitive men, it represented the stability which was induced by the care of the plants and animals which man had domiciled about him.

With every step upward in the organization of society we find that the number and efficiency of these subjugated creatures increases. Our American aborigines in their primitive state commanded only the dog and three or four plants, yet with this scant help they had already won beyond the lowest savagery and were at the threshold of barbarism. In our more civilized societies of to-day we find the products of near a hundred animals and about a thousand plants as elements of commerce, and each year sees some gain in the number of creatures which we make tributary to our desires.

So far as we can discern, the relations of primitive savages to the animal life about them is on the whole more friendly than is that of cultivated men. It is true that the savage looks to the creatures of the wilderness for the greater part of his needs. He slays them, not at all in sport, but for the profit they may afford. Moreover, in most cases, his imagination endows these wild creatures with a spirit like his own. He often adopts them, in his religious worship, placing his tribe under the protection of one or another, as some of our own people do themselves under the protection of particular saints. The effect of domestication when man comes to have [220] his own separate estate in animal life is to separate men from the creatures of the wilderness. "Wild" and "tame" come to be terms having a meaning which the savage does not recognize, and this meaning has with the advance of culture become intensified, until to most men the only creatures entitled to protection are those which have been made subject to man.

At first the process of domestication concerned only useful animals or plants, those which would take a part in our industries. Rapidly, however, these creatures have been adopted with the view to the æsthetic satisfaction which they might afford. Quite half of the number of species which have come under human control have been tamed mainly if not altogether because of the charms which they possess. If we reckon flowering plants in the category, by far the greater number of our captives have been brought to us because of their beauty.

The work of domestication has in the main been effected by our own Aryan race. Out of the total number of animals and plants which have been made captives, probably more than two-thirds have been brought into subjection by the European Aryans or by the folk whom they have profoundly affected with their civilizing motives. The disposition to win goods from the wilderness is in effect a fair test of those qualities in a people which give them dominance: we may indeed roughly measure the qualities of diverse folk by a variety of conquests of this kind, which they have made. The reason for this relation is plain. Success, whether it be of the individual or of the race, depends in large measure upon forethoughtfulness, on a disposition to study as to where profit may be had, and intelligently to [221] seek accessions of strength by experiments in domestication. Each of these winnings from the wilderness represented by our domesticated animals or plants has been painfully and laboriously gained. The men who did the tasks were not creatures of the day, but foresightful beyond the average of mortals.

In a large way the work of domestication represents one of the modes of action of that sympathetic motive which more than any other has been the basis of the highest development of mankind. Ordinary men of the low grade are content to slay, or otherwise rudely gain what value they find in the wild creatures. Only the higher grades of men perceive much of the charm in the inhabitants of the wilderness, or desire to win them to their homes. If our conquests from the wilds were limited to the grossly profitable life alone, we might say that interest only had determined the work of subjugation; but as soon as men escape from their primitive state, even while in their general motives they are still essentially barbarians, they cultivate flowers and derive a keen pleasure from their company. They domesticate birds which are valuable only for the pleasures which their presence lends to human abodes. This action clearly shows that the element of sympathy, that love for the other life which in any way fixes the attention, has had much to do with this work of bringing other beings into association with our own lives.

Not only is the motive which has led our race to such extensive conquests over the wild nature in itself sympathetic, but the process of winning these creatures from the wilderness has served effectively to extend and amplify this same impulse. One of the best features of agricultural life [222] consists in the great amount of care-taking which it imposes upon its followers. The ordinary farmer has to enter into more or less sympathetic relations with half a score of animal species and many kinds of plants. His life, indeed, is devoted to ceaseless friendly relations with these creatures which live or die at his will. In this task his ancient savage impulses are slowly worn away, and in their place comes the enduring kindliness of cultivated men. When we compare the state of mind of the hunter with that of the care-taking soil-tiller, we see the vast scope and influence which this work of domestication has effected in our kind. To it perhaps more than to any other cause we must attribute the civilizable and the civilized state of mind.

Although no discreet person will venture to determine the relative weight which should be given to the influences which have made for civilization, there can be no doubt that the care of domesticated animals has been one of the most potent of these agents. Not only has this employment served to develop the motives of care-taking that result in the postponement of the momentary satisfaction of indolence or of hunger for the prospect of security or wealth to come, but it has served to arouse and broaden the sympathies given men, that humane spirit without which the best of our higher culture cannot be attained. If this view be correct, we may find in it a good reason for regretting the increasing development of cities, a reason which is more definite than the most of those which have been urged against the growth of great towns. Statistics seem to indicate that people are as healthy, as long lived, and on the whole no more given to vice and crime in a well-ordered urban life than they are on the farms. It is certainly easier to give them the formal [223] education of the schools in the dense than in the scattered condition. There can be no doubt, however, that the practically complete separation of the most of our cities from all educative contact with the ancient companions and helpers of men brings about an omission of an element in culture that may entail serious consequences.

The question arises as to what can be done to diminish the evils which come from the total separation of a large part of our people from the humanizing influences due to the care of animals. How general this separation is may be judged from the fact that so far as I have been able to find in the manufacturing towns of Massachusetts not one child in thirty ever knew what it is to care for any creature, save those of its kind. And even in a well-conditioned place like Cambridge, the proportion of those who have any educative contact with animals probably does not exceed one in fifteen. I do not reckon the mere chance playing with a dog or cat as serving the need; the real service is when the person has a sense of responsibility for the life of the animal. To bring about this relation in the ordinary conditions of a town is usually impossible. Something can, however, be accomplished by various expedients.

In the lowest state of townspeople it is out of the question to give the children any pets whatever. Even caged birds cannot or should not be accommodated in the cheaper grade of lodging-houses. Wherever the animals are in separate houses it is often possible for children to have some contact with sympathetic animal life. In these conditions, our cocks and hens are the best creatures to rear. They are the most attractive of all our domesticated birds; they do better than any other forms of economic value in narrow con [224] ditions, and, what is of importance for the end in view, they contribute a share of food, so that a boy may have from them some experience with the economic relation of animals to men.

Some persons who have observed the advancing process of destruction of the natural world may have been brought to consider the change as in the necessary and inevitable order which comes with the higher development of man. They may welcome—indeed, some evidently do welcome—the chance that the ancient system may utterly disappear, and all the earth become fields and garden places tenanted only by those forms that man may have chosen to be his companions. To many people who have a keen impression as to the importance of man in the great economy, and no clear sense of his relation to the natural order, this possibility is doubtless attractive. It is not so to those who have gained a clear idea of the place of man and the conditions of his ongoing.

There is reason to expect that the modern gains in the cheapness and speed of transportation may before long bring about a material change in the housing of the laboring classes of our cities, so that they may be able to dwell in somewhat rural conditions. In this way we may hope to see these people once again brought where they may receive a fuller share of the influences which have served so well to lift our race to its elevated moral station. Working to the same end is the spirit which is leading many manufacturers to place their establishments in the country, where they can control the mode of life of the employees and their families. Against the growth of the factory towns with their sordid conditions, we may with pleasure set these rural [225] workshops where the capitalists are doing the best they can to better the mode of living of the people who are under their charge. In this good work it may well be possible to include a share of contact with the soil and with domesticated animals. In this system of isolated factories we may perhaps hope to find the way out of the perplexities which the present condition of our industries have imposed on our civilization.

Up to our present half-century the process of winning animals and plants to domestication, and of improving them after they had been thus won, has been in its nature a matter of haphazard. Here and there, as men have seen creatures which promised in captivity to afford either pleasure or profit, they have endeavored to convert them to use. In some cases the effort has been made with some patience and steadfastness of purpose. If the creature yielded quickly to the needs of a new life which it was sought to impose upon him, he became a member of man's family. If its wilderness motives were strong, the effort to domesticate was soon abandoned. The greater part of these efforts to win animals and plants into alliance with our race have been made with the creatures which were native in the wildernesses about our ancestral dwelling-places. Occasionally from distant lands important gains have been made, especially among the food-giving plants; but all the animals of any importance which have been adopted by the Aryan people were originally natives of the lands in which that race has dwelt.

It is a remarkable fact that no sooner does a wild animal or plant become intimately associated with man, than it at once departs more or less widely from its ancient type. Our [226] conquests from the vegetable world have to a great extent so far lost their original character that we can no longer determine the species from which they sprang. Botanists cannot find the wild forms which have given us the cabbage, wheat, and most other small grains, and a host of other important varieties. So, too, the origin of our dogs is as yet unsolved and bids fair ever to remain a mystery. In addition to this changed character which we observe in the forms of domesticated animals and plants alike, we note that the mental characteristics of the former undergo vast alterations. The creatures, in a way, take the tone of civilization, and to a great extent abandon those ancient habits of fear and rage which were essential to their life in the wilderness. The intellectual condition of our dogs shows us that the creatures may be progressively educated — in a word, that man may put into them something of his human quality. In the case of the dog, the longest possessed and most familiar to our households of all our captives, the mental change which has come, partly by selection, from association with man has gone so far that the species may be fairly said to have replaced its pristine motives with those which it has derived from ourselves. In many cases it has become, so far as its ways are concerned, even more man than dog.

Although the physical and mental educability of animals when brought into companionship with man is an old subject of remark, and one of the most interesting features which they exhibit, it was not until the doctrine of descent by variation of species from other related forms became established, that we had a chance to see the vast possibilities of accomplishment which are presented to us by our domesticated creat [227] ures. It is true that the breeder's art is old and that men have felt the subjugated animals to be almost like clay in the potter's hands, but except in a small and rather careless way with the dogs, little attention has been given to the development of the intelligence of these captives. The success which we have obtained with this animal has been accomplished by a selective process, but one which has been almost as blind in its operation as the choice which acts in the natural world. For thousands of years men have preferred the dogs which manifested a sympathy with them, and the result is a creature which, though derived from a very brutal ancestry, has in its way as intense affections as human beings. Now and then they have chosen deliberately to develop some mental peculiarity of the animal which would be of service in hunting, and the effect of this care is to be noted in the considerable variety and perfection of mental development which the sporting dogs exhibit. In the main, however, the interest of our dog fanciers has been limited to the physical features of the species; nothing like a deliberate effort to ascertain how far the development of their mental parts could be carried has ever been essayed. In no other field of human endeavor of anything like equal importance has there been so little understanding applied to the tasks.

Now that we are beginning to know something of the laws of inheritance, it is high time for us deliberately to consider what our relations to the organic world are hereafter to be, and how we can guide ourselves in these relations by the light of modern learning. It is in the first place clear that the subjugation of the earth which necessarily accompanies the development of civilization, inevitably tends to sweep away a large part of the organic life which is not [228] adopted and protected by man. Already, with the mere beginnings of this culture, we find that several of the large beasts and birds and a number of plants have been destroyed. New as civilization is on this continent, it has already brought the moose and the buffalo to a point where they are on the verge of extinction, and in the Old World the wild ancestors of the horse and the bull have quite disappeared from the wildernesses. Within a few centuries the greater birds, the Dinornis and Epiornis, as well as the interesting Dodo, have vanished from the southern isles which they inhabited. In the century to come we can foresee that this process of effacement of the ancient life will go on with accelerated velocity.

It seems inevitable that man should play the part of a destroyer. It is his place to break down the ancient order determined by what we call natural forces and in its stead to set a new accord in which the economy of the earth will be in a great measure controlled by his intelligence. Even those who most keenly sympathize with the wilderness life, are not likely to object to the changes which are necessary to open the way for this new dispensation. They may fairly ask, however, that hereafter the displacement of the ancient life shall be brought about with foresight and with the exercise of the utmost care in minimizing the sacrifices which we are called on to make. Naturalists may fairly ask men to remember that each of these species which we are forced to destroy represents the toil and pains of unimaginable ages, and that when these creatures are swept away they can never be recovered. Whatever new species may come, by processes of evolution from the life which remains after we have done our will with the wilderness, we shall never see again the forms which have passed away.

[229] It is the worst feature of the destruction which man is bringing upon the organic species that the assault is most effective on those varieties which are most interesting both from an intellectual and an economic point of view. To take only the case of the great birds which have recently been swept from the earth, we see clearly that we have with them lost precious opportunities for enlarging our understanding of nature and have at the same time been deprived of the chance to domesticate creatures which would most likely have proved of much economic value. With each of these species which disappears we lose what may be a precious chance of adding to the small store of animals or plants which may contribute to the well being of our kind. These considerations make it plain that it is our duty by our civilization, to do all in our power to save these species and at the same time to essay their domestication, for only when under the protection of man can they be regarded as insured from destruction.

The task of bringing wild creatures into our domestic fold is one of very varied difficulty. Many plants are easily reconciled to the conditions of our fields and gardens: they may be said to welcome the care of man which insures them some protection from the fierce contention with other life or with the elements to which they are exposed in their natural conditions. Only here and there is it necessary by careful breeding to develop domesticated habits to the point where the forms will endure culture. Where the task is, however, to make avail of some natural peculiarity which promises to be useful, but is not yet of economic value, it may require a hundred generations of careful selection to develop and fix desirable features. We are, however, in all cases sure in [230] these half-animate species, the plants, that they will prove perfectly obedient to our will. It is otherwise, however, with wild animals. Here we have to deal with intelligences in which the most striking characteristic is an abiding fear of the master, and a general indisposition to submit to any other control than that of their native wild instincts. The measure in which this wilderness habit, bred of long contention with enemies, prevails in animals varies greatly. Some, as for instance the elephant, at once reconcile themselves to human association, and directly on being made slaves accept the mastery of their captors. Others, such as the zebra, remain for a lifetime possessed of their original savage nature. A large part of the labor which has been given to the work of domesticating by the breeder's art the score of mammalian species which man has won to his use has been devoted to this task of expelling the wilderness motives from these forms. The cases in which he has failed to accomplish this end are those in which the savage humor has persisted for so long a time that he has been forced to abandon his effort to subdue the stock.

It seems likely that at the present time we have acquired from the wilderness nearly all the animals which are capable of adoption by such brief and individual experiments as have won to us the species which constitute our flocks and herds. Our future gains will have to be made by far more deliberate and continuous endeavors. These tasks of the hereafter will have to be undertaken in a way which will insure a continuity of effort such as can only be attained by permanently organized associations which may continue their essays if needs be for centuries. The work should be done with two distinct ends in view: first, to determine what members of the wilder [231] ness life may be made to contribute to the needs of man; and, second, how far it is possible so to develop the intelligence of the lower animals in general as to make them better fitted for companionship with our kind. This last-named line of experiments needs to be undertaken not only with reference to varieties now wild, but also upon our most domesticated forms, for, as before remarked, we have not begun to explore the possibilities of intellectual gain, even in those species which have been the longest associated with us.

In considering a list of the creatures which might well be made the subjects of trial with a view to their domestication, we find ourselves at once embarrassed by the exceeding wealth of our opportunities. It is impossible within the limits of this article to treat, even in the catalogue way, a vast number of forms which commend themselves for experiment. Something of the richness of the field, however, may be judged by noting some of the more conspicuous forms, as we shall now proceed to do. Beginning with the insects, the lowest forms in the animal series which have proved in any sense domesticable, we note that wide as is this realm of life it offers but few opportunities such as the domesticator seeks. Of the million or more species in the group, only two, the honey-bee and the silkworm, have been won to man's use, and there is not another wild form which the naturalist can suggest as likely to prove a valuable captive. The only use which we are probably to find for these creatures is where, by some form of culture, we may induce predatory or parasitic species more effectively to do their destructive work on noxious forms of the class. So well fitted is this group for purposes of self-defence that however [232] much man may interfere with the course of nature, he is not likely to sweep any of their multitudinous kinds from the earth, though experience clearly shows that by the methods above mentioned they may be greatly reduced.

It is among the vertebrate forms alone that we find animals which by their characteristics of body or of mind are well fitted to have an economic or social value. There alone are the qualities of flesh or of the external covering such as to make them in a high measure valuable, and the instincts of a nature to fit them for association in man's work. Even among these back-boned animals we find that the lower groups — the fishes, the amphibians, and the reptiles — promise little in the way of gains as compared with the higher groups, the birds and mammals; yet even among these inferior creatures we find certain forms which give promise of improvement under the care of man. Some of the fishes readily learn to come to any one from whom they may expect food, and they indicate in other ways that they are capable of a certain intellectual advance. The frogs and toads readily learn to recognize a master. Several of the larger members of the first-named forms could advantageously be bred so as to be very useful as food. The common hop toad of our gardens is an admirable helper in restraining the excessive development of certain slugs and insects. The tortoises and turtles contain a number of species which are edible, and many of the forms invite the breeder's care. It is, however, when we ascend in the type of vertebrates to the level of the birds that we find the great array of creatures which are worth considering as members of our civilization.

Nearly all the birds except those of prey and those which [233] haunt the seas can easily be accustomed to man. A few of these species which have been reduced to captivity have not become sufficiently reconciled to the unnatural conditions to maintain their breeding habits. Even in these cases, however, it seems likely that in spacious aviaries, at least in climates to which they are accustomed, it will be possible to secure the continuous reproduction of the kind, on which all development by the breeder's art depends.

The ease with which most birds, except those of prey, may be reduced to domestication is due to the remarkable intensity of their sympathetic motives. In this regard the class is much more advanced than that of the mammalia to which we ourselves belong. Accustomed as they are to ceaseless and active intercourse with each other by means of their varied calls, largely endowed with the faculty of attention, and provided with fairly retentive memories, the birds are, on the average, nearer in the qualities of their intelligence to man than are many of the species in his own class. It was long ago remarked that the birds of remote islands, such as the Galapagos, which had never seen man, were at first not in the least afraid of him. It required, however, but a few generations of experience to show these creatures that the unfeathered biped was a singularly dangerous animal, and they at once and permanently adopted the habit of avoiding him. This incident of itself shows how quick birds are to learn certain large and important lessons. We see also the reverse of this education in fear in the rapid way in which birds become tame when they are secured from persecution. Wherever shooting is stopped over a considerable territory the birds rapidly become more tolerant of man's presence. Even among migratory species the individuals [234] appear to learn that certain places where they are protected may be resorted to with safety.

Because of their freedom of flight it is in all cases difficult to bring our perching birds into such relations with the domiciles of man that they can be truly domesticated. The success, however, which has been attained in the case of the pigeons, which have been so far made captive by the change of their instincts that they never depart far from their cotes, seems to indicate that this tendency again to go wild is by no means ineradicable. In other instances it will probably disappear as it has in this by long-continued care in breeding. Our successes with the geese, ducks, and swans, all of which belong to genera characterized by the migratory habit, show how readily in the course of time the old native instincts may be subordinated to the will of man. Although the degree of the difficulty which will be encountered in taming many wild forms may be far greater than that which has been met in those which we have domesticated, there is no reason whatever to believe that in any case it will be insuperable.

While all the creatures of the wilderness may by the breeder's art be induced to vary in the conditions of captivity, the birds have shown themselves more plastic in our hands than any other animals. In almost every brood we find individuals possessing marked peculiarities of form or plumage. In their mental qualities also there is a like range of variation. Seizing upon these, the fancier can guide the quick succeeding generations so as to cause the form to depart in the course of a few years very far from its original aspect. With each step in this succession of changes the readiness with which the species responds to selective care increases. The results which have been attained in our barnyard fowl [235] and with the pigeons show how admirably these creatures are fitted to obey the will of man when he has a mind to take charge of their destiny.

Perhaps the greatest conquests which we have yet to make among the birds will be won from the species which have the habit of dwelling mainly or altogether upon the ground. These, as experience shows, can be more readily brought to the uses of man than the species which are free by their strong wings to wander through the realms of air. There are very many of these ground birds the domestication of which has never been fairly essayed. There are perhaps a hundred species which in one part of the world or another might afford valuable additions to our resources, those of ornament or of economy, and yet within three centuries only one of these, the turkey, has been brought to the domesticated state. The greater part of our game birds, such as the quail, pheasants, and partridges, though they appear on slight experiments to be untamable, could probably by continuous effort be reduced to perfect domestication. For ages they have been harried by man in a manner which has insured a great fear of his presence. We have indeed through our hunting instituted a very thorough-going and continuous system of selection which has tended to affirm in these creatures an intense fear of our kind. Only the more timorous have escaped us, and year after year we proceed to remove with the gun the individuals which by chance are born with any considerable share of the primitive tolerance of man's presence. It is not to be expected that the chicks of these species will at once accept relations with our kind. The domestication of many of these forms is to be desired, not only on account of the excellent quality of their flesh, but because of [236] their beauty and the charm which their quick intelligences afford them. Whoever has watched them in their care of their young or their other social habits has observed features which indicate a possible development under domestication perhaps greater than that which we have attained in any other of our feathered captives.

It seems most important that experiments in the further domestication of birds should be first addressed to certain, large ground forms which are now in more or less danger of extinction. The newly instituted industry of ostrich farming has probably insured this the noblest remnant of the old avian life from destruction; but the emu and the cassowary are still among the diminishing and endangered forms which unless taken into the human fold are likely soon to pass away. The brush turkey and the bower bird of Australia, two of the most curious inhabitants of that realm of strange life, appear to have qualities of mind and body which would make them readily domesticable and which would cause them to be among the most interesting of our feathered captives.

Among the aquatic birds there are many species which are as promising subjects for domestication as any which have been made captive; these if subjugated would prove great additions to our resources of ornament and use. Thus the eider duck, so well known for its wonderful soft down which is plucked from the breast to make a covering for the eggs, though a marine species, would prove domesticable at least on the seashore of high latitudes. There are many other varieties of the family, such as the canvas-back which is so highly esteemed for its flesh, that would likewise afford very interesting subjects for experiment.

The wading birds are characteristically very wild and range [237] over a wide field; yet the flamingoes, the herons, and their kindred could probably be brought into at least as near an approach to reconciliation with man as their relations the storks. The comfortable relations which have been established between the last-named species and humankind in northern Europe is probably in nowise due to the peculiarly tamable nature of the bird, but rather to the fact that certain superstitious fancies on the part of the featherless biped led him to protect the feathered visitor of his roofs and chimneys. Should it be desirable to break up the habit of migration in these or other birds which are now accustomed to range up and down the meridians, there seems no reason to doubt that the change could be accomplished with the same ease that it has been in the case of the tamed geese and swans. Experience has shown that with these forms, which probably have not been associated with men for more than three or four thousand years, the migratory instinct, which appears one of the strongest of motives, has utterly disappeared. Not only do they no longer heed the cries of the wild birds of their kind as they fly away on their annual journeys, but they have, through the changes in form induced by their quiet life, lost the power to rise far above the earth. They are even more effectively tamed than are their captors.

Owing to their singularly perfect protection against the cold, and also perhaps to the quickness of their wits, birds are more readily transferable from one clime to another than are any other animals. The feathered tenants of our barnyards are, except perhaps the aquatic species and the turkey, all from the tropical realm. Experiments with various other wild forms go to show that there are very many other [238] tropical species which will prove to have an equal tolerance of high latitudes. If this be true we may fairly look to the domestication of the varied bird life of the equatorial regions for the enrichment of our northern lands. Even when it may not be desirable to bring these species to the state of complete subjugation they may be introduced on something like the terms which have been given and accepted in the case of the so-called English pheasant, which has brought to the high north of Britain and some parts of this country an element of grace which is afforded by no indigenous form of North America or Europe. There are hundreds of beautiful tropical species which await reconciliation with men; they have that quality of sympathy which affords the natural foundations for the contract, but this has in no case been availed of except when the creatures, in addition to their æsthetic charm, have possessed some economic value. There as elsewhere in the matter of domestication the commercial motive has controlled our action.

In forming our societies as we are in time to do, account must be taken of the sympathetic value of its elements, reckoning among these the animals which the system brings in contact with men. Much of the culture which has served to lift our race above its ancient savagery has been derived from the influence of domesticated animals; in proportion as these creatures have sympathetically responded to our care we have been thereby educated and our spiritual development advanced. So far as in our further choice of animals which are to be associated with ourselves we are guided by a desire to extend this work, we may well turn our attention towards the birds, for in that group we may find a greater number of species which have attained the physical [239] beauty which attracts and the mental qualities which may endear them to mankind. They can give us nothing that can ever come so close to us as the dog—the unique gift of the wilderness—but they may afford a host of forms to enrich our lives.

The mammals, because they are, in qualities of body and mind, nearer to us than the members of any other class of animals, afford the most promising field from which to make selections for future domestication. In an economic sense it seems unlikely that any very great profit can be attained by the subjugation of any of the mammalian species which are still wild. Civilized people have been so long in contact with the life of all the continents, and have ever been so hungry for gain, that they have already essayed about every experiment in subjugating the larger wild beasts which appears to be very promising. Still there are certain cases where there have been no trials and others where the failure to tame particular species has been due to hindrances which systematic labor may possibly overcome. It will therefore be well to glance at the array of the wild forms which afford some prospect of success in the hereafter, including under the title of successes those kinds which may contribute not only to immediately measurable wealth, but the æsthetic satisfactions as well.

Beginning with the lowlier group of mammals we find in the base of the series the ornithorhynchus and its allies, creatures which have nothing to recommend them but their exceeding organic peculiarities that render them attractive to the naturalist, but which are not likely to win them a place in the affections of men in general. As these species are most inoffensive as well as interesting, and as they [240] are now confined to a portion of Australia, they might well be made the subject of some human care which would stop short of domestication. They might be transplanted to other continents and thereby given a larger field for variation as well as a chance to exhibit their features in a wider field. Among the pouched mammals, especially in the species of kangaroo, there are forms which commend themselves as very fair subjects for taming. They are of considerable size, their flesh is palatable, and their hides useful for leather; they breed rapidly, live on a poor herbage, and are, for wild animals of like strength, very inoffensive. Moreover, though relatively invariable both in mind and body, they exhibit sufficient individual peculiarities to indicate that the breeder's art could, in a short time, bring about considerable changes such as have been effected in other species, changes that would increase the value of these animals. As far as æsthetic or sympathetic relations are concerned, the pouched mammals have nothing to give us; they are, as befits their lowly estate, among the least graceful of their class; they are also little interesting in their mental qualities, being about the stupidest of our kindred.

Among the ordinary herbivorous mammals there are several which should be domesticable which have not yet been properly subjected to experiment looking to that end. The American bison, commonly but improperly termed the buffalo, is a strong creature, one which is easily nourished. In its present condition, it is about as promising a subject for the breeder's care as were the ancestors of our horned cattle. Although there have been sundry trials of this animal as a beast of burthen, they have been of a rude as well as a brief kind, no care having been taken by selection to improve [241] the qualities which evidently commend themselves to our use. The flesh of this species is quite as good as that of the wild bulls of the genus Bos, and the hides have a peculiar value on account of their somewhat woolly character. There is reason to believe that, bred in the region of the high north, about Lake Saskatchewan for instance, with proper selection this hairy covering could be developed much as has the wool on the sheep. This is indicated by the considerable variations in the quality of the coat which go to show that the feature is still in a very plastic state, a state that may be said to invite the assistance of man in order to bring it to the full measure of its possibilities. If this covering could be developed, the result would be to give us a domesticated beast of large size with a hairy covering having the character of a fur; such would be a great addition to our resources.

As there is a large extent of country in the high latitudes of North America, Asia, and South America, where the climate is too severe and the herbage too scanty to serve the needs of our ordinary cattle, in which a hardy feeder with a well-clad body such as the buffalo might do well, it seems most desirable to essay the experiment of domesticating the bison before it is too late, before the brutal instincts of our kind have quite made an end of the noblest animal which is native in the Americas.

There is another inhabitant of the high north of this continent which deserves the notice of those who are disposed to attend to the questions concerning the extension of man's control over nature; this is the ovibos or musk-ox. Like the buffalo, only in much higher measure, this singular creature is fit for very cold countries; his fitness being in [242] part assured by his admirable covering of long hair as well as by his capacity for taking on fat during the short summer in sufficient store to last him through the trials of the winter season. The kinship of the musk-ox to the group of the sheep is near enough to warrant the belief that the hair could be improved by selection, and that from the process we would be likely to obtain an animal much larger than our largest sheep and yielding fleeces of peculiar value in the arts.

Among the northern carnivora there are several species which deserve attention for the reason that they may be brought to some degree of domestication which may enable us to make better use of their hairy coverings. Among these we may mention the foxes, the polar bears, and the seals. The first-named group affords at present about the dearest furs of our markets. The silver-gray variety, which at present seems to be a frequent individual variation, could doubtless be affirmed by selection, and probably could be brought to a higher state of perfection than it has as yet attained. The animals are, if we may judge from their kindred, not untamable; at least they could be brought to live in a sufficient state of captivity to permit selection. In time they might be quite domesticated. Many of the islands of the high north and south are well fitted for such experiments.

As is well known, the polar bears have a wonderfully developed hairy covering; their coats, indeed, are among the richest that exist. These animals subsist mainly on what they capture from the sea, so that it might be possible to keep them at a small expense. They are, however, of all their kindred the most indomitable; it would probably [243] require a long and costly effort to reduce them to anything like domestication. Moreover, being strong, free swimmers, it would not be easy to maintain them in captivity. Still, selecting such a well-inundated place as Bear Island of the North Atlantic, it would be most interesting to make the experiment, first of accustoming them to some human control, and then to a selection which might serve to lift the quality of the kind. It would be less difficult and perhaps more advisable at first to make a trial of a similar sort with the black bear, which in less arctic conditions flourishes and carries a fine pelt. The only difficulty would be in finding a sufficient supply of food for such captives, for although they will eat fish they have no skill in capturing them such as is possessed by their more degraded, or perhaps we should say their less advanced kindred, the polar bears. Still, as the form is even more omnivorous than man, it might be practicable to feed them.

By far the most important of the carnivora in an economic sense are the seals which dwell in the high northern waters. These creatures afford the most interesting subjects for experiments in domestication from an economic point of view that remain to be made. Of all the predatory animals the seals seem to have the largest share of intelligence and the greatest amount of sympathetic motives. No other wild animals, except perhaps the monkeys, appear to be so human-like in their qualities of mind as these creatures of the sea. So far, except when they have been captured and kept for purposes of show in menageries, man's relations to the seals have been purely destructive; he has incessantly hunted them. Yet certain species of them remain singularly willing, we may say desirous, of claiming friendship with their persecutors. As [244] elsewhere noted, wounded seals behave in a curiously appealing way towards their assailants. When in captivity certain of the species show a remarkable friendliness and a capacity to receive training. No other wild animals, except perhaps the elephants, exhibit so great a fitness for profiting from contact with man.

Although our knowledge as to the habits of seals is still very imperfect, it appears likely that the greater part of the species have the habit of resorting to certain places during the breeding season, and that the individuals after the manner of certain fishes return at that time to their native shore. If this be true, as there is good reason to believe it is, it should not be a matter of grave difficulty, provided the maritime nations would abet the experiment, to establish seal colonies composed of the several promising forms at fit points in the circumpolar seas. There is reason to suppose that with ordinary decent treatment the animals would become to a great degree accustomed to men, and that it might be possible to accomplish selection enough of the individuals which were left to breed, to develop the already valuable characteristics of the fur. In the present disgraceful condition of our relations to these animals it will be but a few years before we shall have to lament the extirpation of several species, including the most interesting members of the group.

Looking upon the questions of man's future on the earth in a large way, we see that there are reasons why the animals of the high north, particularly those which obtain their food from the sea, should be protected from extermination. There is a great area of country in that part of the world which is not adapted to the occupation of any of the species which have as yet been domesticated. If this portion of the world is ever [245] to prove fruitful in other ways than through its mineral stores, it will be by the creatures which are adapted to its climate and other conditions. At the present rate of increase in numbers, the population of the world will, in the course of two or three centuries, begin seriously to press upon the resources in the way of food which the fields of the tropical and temperate zones can supply; the chances of the arctic regions may then have much importance to our successors. Moreover, in the case of the seals we find the peculiar advantage that the animals are fed entirely from the sea, so that the domestication of these forms would give to man a means, the like of which he has never possessed, whereby he would be enabled to harvest the food resources of the deep.

The beaver, particularly the North American form, offers a most attractive opportunity for a great and far-reaching experiment in domestication. On this continent, at least, the creature exhibits a range of attractive qualities which is exceeded by none other in the whole range of the lower mammalian life. No other mammal below man shows anything like the same constructive skill in the contrivance of its habitations, or is able so to modify its habits of building to meet the varied needs of its life. When this country was first visited by man near one half of its area was occupied by this species. It built its dams and dwelling-places and, when necessary, excavated its canals along all the lesser streams in the timbered regions of the northern districts. As the destructive effects of civilization increased, the animal has gradually, to a great extent, been driven away from its old haunts, and where it remains it has, as the price of life, given up its architectural habits and betaken itself to the older and simpler mode of living in a chance manner much as is now [246] the habit of the European variety. As an illustration of this I may note, in passing, that before the civil war, when all the recesses of the forests in the region about Richmond, Virginia, had for more than a century been industriously explored by hunters, the beaver was supposed to be extinct in the district; yet during the civil war, as I am credibly informed, a colony of these creatures became established near the town of Suffolk, and there, amid the roar of a great conflict in which men ceased to seek the lesser game, they recovered their habit of building dams, which we must believe to have been discontinued for many generations. This capacity to vary action with reference to changing needs is the best possible index of the mental power of animals. Guided by the exhibition that has been given us by the beavers, we are justified in considering them to be the one group of mammals which has gained a distinct, rational constructive power. This feature makes them decidedly the most interesting group for investigations which may be expected to throw light on the problems of animal intelligence. From the economic point of view the species has a certain importance for the reason that it affords one of the most valuable kinds of fur that has ever been marketed.

The domestication of the beavers to the point where they would tolerate the presence of man should not, provided they could be protected against the depredations of poachers, be a matter of any difficulty. The colonies of these animals require only what is afforded by vast realms of our wildernesses—flowing streams of moderate fall with timber upon their banks. They are not particular as to the species, so that swift-growing kinds of trees such as the poplars may [247] be made to serve their needs. The natural growth on a hundred acres of otherwise worthless land would probably be sufficient to maintain a colony of average size containing say twenty-five individuals. In the region about the great lakes and for some distance to the northward and to the east and west there are great areas amounting in the aggregate to some hundred thousand square miles that would apparently be well suited to the nurture of this form, and which in the present condition of the country, as well as for the immediate future, cannot be turned to better use. It may be remarked that the domestication of the beavers would afford yet another means, in addition to those above noted, whereby we might be able to win some profit from the great wilderness of the north, which is, so far as our existing means of appropriating its resources, of little use to mankind. The only evident way by which we may hope to win profit from this part of our continent is by using it as a field for rearing animals that have yet to be subjugated; none of our captive varieties are fit for the service.

In the tropical parts of the world there are many mammalian species which are worthy subjects for essays in domestication. This is particularly the case in the continent of Africa where, except in the lands about the Mediterranean and the Red Sea, the native peoples have never attained the stage of culture in which men become strongly inclined to subjugate wild animals. Africa is richer in large herbivorous species than any other of the great lands; many of these forms are of large size and have qualities of flesh, of hide, or other peculiar features which promise to make them valuable in an economic way. Others, especially the antelopes, have a beauty of form and a grace of movement which [248] render them among the most attractive creatures of their class. Even the hippopotamus, one of the grossest beasts of this realm, affords in its teeth a valuable ivory, and its hides, if supplied in sufficient quantity, would probably find a considerable use. It is evident that in this "dark continent," where the influences which make for human advancement have been so slight, we have the best field for the selection of species that may hereafter be brought to the use of man. There is evidently danger, in the advance in the civilizing process, that the native forms which, owing to their fitness to the physical conditions of the country, might be made useful to its people, may be utterly destroyed by hunters.

Perhaps the most interesting of the tropical beasts from the point of view which we occupy is the elephant: This animal in its relations to men is eminently peculiar, in that while it has been in an individual way long and completely subjugated, it has never been systematically reared in captivity. Owing, it may be, to the slow growth of these great beasts, as well as to the immediate manner in which they submit to their captors, it has ever been the custom to take them when adult from the wilderness. The result is that the supply of the Asiatic species, which alone is serviceable—the African form being apparently too fierce for use—is now dependent on a relatively small number of wild herds. Certain of these herds are protected by the governments of India, but it seems as if the species were already dangerously near the vanishing point—in a position where the invasion of some disease or some insect enemy might deprive the world of what is, all things considered, the most interesting of the brutes. Moreover, the failure to rear elephants in captivity has made it impossible to essay any of those experiments in breeding [249] which have done so much to improve the utility and the beauty of most subjugated forms.

If the elephants could be reared in captivity there is little reason to doubt that with a few centuries of selection they might be made to vary in many important ways. It is evident that the form and mental quality of these creatures is as plastic as those features in the other domesticated animals have been proved to be. Moreover, the group, though it is now represented by but two recognized species, was in comparatively recent times quite rich in varieties, a fact which raises the presumption that the existing kinds are open to modification by the selective process. As the elephant is not mature until it is near thirty years old, probably not reproducing until about that age, there is little inducement for any person to undertake the process of breeding them in the selective way; if the task is ever done it will have to be accomplished by government action or by that of a society which is pledged to such tasks. If the effort to bring the elephants into a more permanent relation with man is not made and the race is allowed to perish, we may be sure that in the time to come people will gravely censure us for any such neglect of the opportunities which this world affords as would be involved in the loss of this noble brute. It is clearly our duty to see that all such resources are preserved for the inquirers of the future.

Among the other tropical mammals which, because they have not as yet proved of economic value, are on account of their size and their attractiveness to sportsmen in danger of extinction, we may note the various species of rhinoceros, the giraffe, and the several African forms which are akin to the horse. None of these forms have been turned to use, [250] none of them appear likely to be adopted by man for the service they can do; but they are, in common with all the host which cannot be mentioned here, of great interest to the naturalists of our time. Their importance in the inquiries which are hereafter to be made by our ever expanding science of life cannot be estimated. It certainly will not be possible to overreckon it in this very practical age. This plea for the sparing of the mammalian species in no case needs to be made so strongly, and in no other instance is so well entitled to a hearing, as when it is raised for the life of the monkeys. These interesting animals because of their collateral kinship with man afford precious evidence as to the stages of intellectual development which is likely to be of exceeding value to students in that field of inquiry. There is unfortunately little chance that any of the monkeys will ever prove useful; their habits are such that they are generally troublesome neighbors; moreover, their weakness makes it easy to exterminate them. The result is that some species have probably already been destroyed, and others are in conditions where during the next century they are likely to vanish. In the animate realm it is hard to choose the forms which are to be the most important for the naturalists of the time to come, but it is certain that these students will deplore the loss of the simian life and charge us sorely if we neglect due effort for its preservation.

Although the matter before us concerns the domestication of animals, it may be well to devote a little attention to the question of the wild plants which need protection or which promise to afford unwon values. It may be said that plants in general are much less likely than animals to be disturbed by the process of bringing a country under the condi [251] tions of civilization. With rare exceptions the individuals of each species are so numerous that, like the insects, they escape by their numbers the risk of the extinction of their kinds. Moreover, the ease with which nearly all the kinds can be brought under cultivation, and the fact that they present no self-will to be dominated, makes the task of dealing with them, in a protective way, infinitely easier than in the case of animals. So far as we know, there has not been an instance in which a continental species of plant has been exterminated by man, while there are a number of the larger animals which have been swept away apparently by human agency, and there are many more which are on the verge of extinction. Therefore, so far as the plant world is concerned, we may for the present at least trust the species to their own powers to maintain them against the rude assaults of civilization. If here and there one is overrun by the wheels of our economic engines, something of value to the student is lost, but the loss does not include the element of mind which is hereafter to be the subject of so much study.

The foregoing considerations make it evident that the problem of domestication shades into the question as to the preservation of the life which is now on the earth, and this with a view to the advantage which the arts, the sciences, or general culture may obtain from the preservation of the useful, the instructive, and the beautiful things in the realm of nature from the swift destruction which our rude subjugation of the earth threatens to inflict. To deal with this problem in an adequate manner we must ask ourselves what limits are to be set to the displacement of the ancient order which is now going on. We see that wherever civilization enters, and [252] even where its first influences are felt, the olden societies of nature are disturbed or broken up. All the nobler members of these associations, the greater mammals, many of the larger birds, and a host of the lesser forms, are expelled or destroyed. In the condition of organic life when the supremely predatory creature man rose to domination, the species were grouped in those vast organizations which were of old termed faunæ and floræ, but which are now better known as biological fields or provinces. In each of these hosts the several species were, as regards their external life, so balanced with their neighbors that the assemblage from the point of view of these relations might well be compared with the polities or states of man's construction. Such an organic society represents the result of a series of trials and balances which began to be made in the immeasurably remote past and have been continued through the geologic ages, each age adding something to the accord. The plants give and take from the animals; the insects are equated with the birds, and each species in every group has set up an accord with its rivals. From time to time the host has by the changes of sea and land been compelled to migrate, moving this way and that to find its fit station. In these movements species are rapidly extinguished, much as the weaker soldiers of an army perish in forced marches. Into their places new forms hasten to take their place, so that every position of advantage is filled. At a less rapid rate, but perpetually, even without the change of abode, which it is often by climatic changes compelled to make, the organic host is slowly changing in character; old kinds give way in the endless contest to new varieties which have managed to establish a better relation to the environment. Still the [253] legions press on towards the great accomplishment of a higher and nobler life.

No one, however well he may conceive the nature and history of the organic hosts of the earth, can hope to convey to the general reader an adequate sense of their majesty or the wonderful part they have played in the history of the life which has culminated in mankind. The largest words are freighted with too little meaning, and even the metaphors drawn from human associations fail to convey a sufficient picture of these enduring organizations which have enabled living beings to meet the difficulties of their long contest with this rude world, and to win the advance they have gained. The reader will have to tax his imagination to picture, it may be, a quarter of a million species dwelling in the same field, each united with the other in the method of exchange in such a way that the withdrawal of any one form is likely in some measure to change the estate of every other. In some cases this removal of one species means the loss of the life of many and perhaps the better opportunity of other neighbors; again, the influence on remoter members of the society may be so slight as to escape detection. Yet it is doubtful if the slightest change in the population of a biologic province can be brought about without some effect upon all the members of the society. It is a vast, sensitive thing, fit to be compared with the living body where every cell lives in accord with every other of the frame.

So long as the organic hosts were in the prehuman stage the maintenance of the accord was easily and naturally attained. Species arose and perished, each in turn effecting a simple reconciliation with the others, grasping only so much room and food as was necessary for its proper support. But [254] with the coming of man, the species which by its swiftly progressive desires has become a host in itself, a disturbing element was introduced into the old order. Man as a primitive savage falls into the natural system without greatly disturbing it; but man as a soil-tiller, in so far as he carries out his subjugative work, utterly wrecks the ancient establishments of life. To attain his object he has to banish from the soil nearly all the plants which originally belonged upon it, and in their place, with or without intention, he introduces species from other organic provinces. With the change in plant-life necessarily goes a like, or even a greater, alteration in the native animals. They are driven into the wilderness or, it may be, extirpated. The reader who would obtain an idea of these changes will do well to study the invasions of weeds or of those noxious insects which in the economy of a civilized country may be likened to weeds. These pests are in nearly all cases invaders which owe their successes to the fact that our treatment of the regions they have entered has opened vacancies in the once closed ranks of the indigenous host, into which the foreigners are free to enter. In the fresh field they are not likely to find enemies which by long training are especially fitted to cope with them, and so they run riot and contest with man the gains he has won from the ancient possessors of the land.

Of all the large questions which the consideration of the future of man's work on this planet opens to us, there is none which now appears to be more serious or, in its consequences, more far-reaching than this concerning the treatment which he is to give to the old natural order of sea and land. The very first condition of civilization is an utter spoiling of that order, so far as the land areas are concerned, in the fields of [255] the richest and highest life. It is clearly impossible to avoid this destruction over all the surface which we win to culture. Spare as we may, the subversion of the ancient balances and adjustments must be complete before the earth is ready for our tillage and other modes of use. This overturning is a part of the destiny of man. It is a characteristic of the new dispensation which came with his progressive desires. Yet the rational quality which has led to the mastery of man may be trusted to bring him to a point where he will endeavor to minimize the ill effects of his actions on the life which has been placed in his hands.

In considering the ways in which we can mitigate the evils of our rule over organic nature, we at once see that our aim should be to preserve all the varieties of living creatures from destruction, provided they are not distinctly harmful to man, and this with the intention of keeping for our successors in the inheritance all that can in any way afford a foundation for further experiments in domestication, materials for learned inquiries, or pleasure in contemplation. To attain this object we cannot trust to the share of this life which can be brought into zoölogical and botanical gardens, however extensive and well managed. The only way is to make certain reservations in various parts of the world, each containing an area and a variety of conditions great enough to afford a safe lodgment for a true sample of the life of an organic province. Owing to the fact that these provinces are never sharply bounded, it would naturally be impossible to select reservations which would in a complete manner represent all the conditions of the biologic societies; but if properly distributed the outlying animals and plants could in most if not all cases be introduced into one or other of [256] these protected fields, so that there would be little reason to fear that any important part of the existing life would be lost.

Owing to the wise forethought of our American people, a practical foundation of the system of national reservations has been instituted in our so-called national parks. Although these reservations were established to preserve to the public certain natural beauties in the way of scenery or vegetation, or to secure the regimen of streams, they will, if properly guarded against depredations, effect the end which we have in view. Owing to their large area and somewhat varied positions, these parks provide a safe refuge for a great part of the life which belongs in the Cordilleran district of the United States. If the method should be extended to the whole country, we should have the peculiar satisfaction of having been the first state to institute the system of preservation which is here suggested.

To complete a system of reservations designed to perpetuate the aboriginal life of this country would require the institution of about a dozen other similar natural shelters. It would not be necessary to have these on as large a scale as that of the Yellowstone. In most cases areas of from ten to twenty thousand acres in extent would, if well guarded, suffice to give refuge to the animals and plants of the field in which it lay. The selection of these refuges would demand much consideration. In general, it may be said that they need to include at least two on the Atlantic coast, which might also be fitted for the use of marine birds as breeding places, one on the northern part of the coast of Maine, and another in southern Florida. The latter might serve as well for the protection of the turtles which resort to that shore to lay [257] their eggs. Similar coast parks should be established on the shores of the Pacific. Yet other closed areas would be needed in the interior, the evidently desirable fields lying in the region about the headwaters of the Mississippi, in the Adirondacks, in the mountains of North Carolina, in the lower part of the Mississippi delta, in Arizona, and at least two points in Alaska; one of these should afford a place of refuge for the persecuted fur seals and another for the musk-ox.

At first sight it may seem to be a simple matter to accommodate the wild life of a country on a relatively small piece of land. So far, indeed, as the plants, the insects, and the lesser mammalian life are concerned, an area of a few hundred acres will serve very well for their safe harborage, but when it comes to protecting the larger birds and mammals we see how easily the natural balance of life is by some chance influence destroyed. A capital instance of this difficulty which arises when preservation is essayed on small areas has recently been forced on my attention. In Dukes County, Massachusetts, there is the vanishing remnant of an interesting bird known from the island to which it is limited as the Martha's Vineyard prairie chicken. It is closely related to its better known Western kinsman, yet is a distinct variety. Although the form has apparently developed on the island and once abounded there, it has dwindled in numbers until there are but few surviving. In the hope of providing a safe refuge for the remnant, I have for a number of years stopped all shooting on a tract of a thousand or two acres which is well fitted to supply them with food and shelter. As they still dwindled, it seemed probable that the foxes were harming them. This appeared the more likely for the reason that the fox is not a native of the island, but was introduced a few [258] years ago by some reckless experimenters. These marauders were cleared away without good results. Further inquiry made it apparent that the real enemy of these birds was the feralized domestic cat which has gone wild from the households, especially from the many homesteads that have been abandoned. This creature has bred in great numbers and is now threatening the existence of all birds that rear their broods upon the ground. It is hardly possible to exterminate them, for the reason that they are wary, and any systematic hunting of them would prove exceedingly disturbing to the very timid birds. The result is that nearly all these birds have left my land for certain plains near by which are covered with scrub oaks and where there is too little ground life to attract the cats. In that region, though it has an area of about thirty thousand acres, the food is scanty; the prairie chickens dwelling there are likely to perish for lack of the rose-hips which, in the hill country they have been forced to desert, served to maintain them at times when the ground was covered with snow.

The lesson which may be drawn from the experience above stated is to the effect that it is necessary to have a protected field of sufficient area, and in the proper conditions to keep the balance of life which arises from the exchange of relations between species in their normal state. Even in ideal reservations where all invasions are excluded, we should have to expect that from time to time certain forms would disappear, their place perhaps being taken by new species which would arise. Such is the manner of the great procession of life. Probably at least twenty and perhaps a hundred times as many species as are now living on the earth have perished from it, and before the unimaginable goal is attained [259] as many others may pass away. Our task with the refuges would be to keep the death of the specific inhabitants to the natural and wholesome rate that is determined by the endless struggle for existence.

It is impracticable at the present time to devise a scheme for refuge stations in other countries than our own; it is evident, however, that these would have to be numerous and widely distributed. A glance at a map showing the political distribution of the lands will make it evident, however, that within the holdings of the British, French, German, Dutch, and Russian governments there are large areas which might, without evident loss of considerable economic values, immediate or prospective, be turned to such uses, and that these reservations would probably include nearly all that would be required to preserve the most important samples of the primitive life. Some of them, as for instance those intended to retain the large tropical animals in their natural state, would have to be as imperial in their areas as the Yellowstone Park, but these would lie in realms which have no present value to our own race and are scantily inhabited by the indigenous peoples.

It is easy to see that the proposed world-wide system of wilderness stations in which the native life should be preserved from the destructive influences of man's assault upon it could not be brought about without international coöperation and with a considerable expenditure of money both for the foundation and maintenance of the establishments; but, as before remarked, the idea of public reservations of this nature is one which immediately and strongly commended itself to the people of this country and has led their representatives to set aside for such use lands which in the aggregate amount to a larger area [260] than some of our sister states. The same motive is seen in the action of the State of Massachusetts, which a few years ago created a Board of Trustees of Public Reservations, a corporate body authorized to hold in perpetuity lands which are intended to serve the public for pleasure and instruction. The recent rapid extension of the park systems appertaining to the cities of this country and Europe is a further illustration of the same motive which makes for the object which we desire. It therefore seems not unreasonable to hope that very soon we may find the governments of the greater nations willing to go forward on the line of advance in which our own has so well led the way. At the right time the United States could probably do much to further the matter by asking for international action in this admirable work. There is hardly any undertaking which would afford a fairer chance for friendly coöperation among the great states than this which looks forward to the good of the time to come.

While looking forward to the establishment of a system of sanctuaries which may serve to protect examples of the present life of all the lands, it is also well to consider what can be done by local authorities and by individuals in the same direction. The numerous zoölogical and botanical gardens which have been established in different parts of the world have in part the same motive that is to be embodied in the larger institutions which we would see founded; they seek to preserve the interesting and instructive animals and plants, and in some cases contrive to perpetuate the kinds. The trouble is that their main purpose is to make a striking show, one that will attract the eye and lead to profit of an immediate kind. If these institutions could be persuaded to add to their former exhibitions grounds designed for the maintenance [261] of the natural order, true wildernesses, where the native life would find a fit place of abode and where it would be protected from the ravages of man or from accident, a certain gain would be made; at least the masses of our city people, who have now come to control legislation in the great states, would be brought to see the beauties of the primitive conditions which they now rarely have a chance to behold. Yet more might be accomplished if men of wealth could be induced to turn their generous spirit towards this object. There are many parts of this country where reservations are most desirable and where the price of land is so low that an area of thirty thousand acres could be acquired for that number of dollars. A capital of one hundred thousand dollars would, at the present rates of interest, afford the revenue necessary for the pay of a keeper and half a dozen guards, a sufficient force to maintain a due watchfulness against depredations. Moreover, the use of such land as an asylum would not prevent a careful exploitation of its timber resources, which in many cases would give a sufficient return to provide for the policing expenses, as well as for incidental costs incurred in bringing upon the land species from the neighboring country which it might be desirable to introduce. At a cost of not more than a million dollars it would be possible to secure and maintain a well-chosen system of guarded wildernesses which would preserve the characteristics of the original plant and animal life in all the region of this country lying to the east of the Rocky Mountains.

It would be essential in any such privately founded system of wilderness reservations to have the control of the establishments in the hands of some authorities which were of an enduring nature. In our American experience it has become [262] certain that such trusts cannot be safely reposed in the state or national governments, or in the hands of trustees chosen for the particular function. The only authorities which commend themselves for the execution of such a purpose are those of our universities. In these institutions we find boards which are chosen for the attainment of intellectual ends; in certain cases the choice is made by the vote of an intelligent body of alumni, or in other ways guarded by that body, so that the chance of lapse in the quality of the contract is reduced to a minimum. Several instances could be given showing that such trusts, even when they do not directly pertain to the teaching work of these institutions, have been long and faithfully maintained. We may therefore look upon our universities as the natural repositories of confidences which pertain to the continuous intellectual work of man. There is no other kind of association where interests of the sort which would have to be cared for in the reservations of the wilderness are so likely to receive continuous attention. In these homes of learning, while business considerations enter, personal greed is naturally absent.

The method which may be chosen for the control of wilderness reservations, though a problem of much importance, is of course secondary to the matter of their establishment. This work should at once command the attention of those persons who are of the foresightful class who see beyond the interests of the day, and take account of the needs of the generations to come. Such men will do well to begin the work by organizing a society which shall endeavor to arouse public attention to the destructive effects of man's occupation of the earth by his civilizations. The people need to be taught the true meaning of the indigenous life in [263] relation to the problems of the origin and destiny of our own and other life, to the future exercise of the domesticating art and to the most refined gratifications.

It may be noted that, beginning with the apparently simple and eminently popular questions as to the origin and economic history of the animals which have been subjugated by man, we have been naturally led to the consideration of much larger problems, those relating to the place of man in the order of nature, and his duty by the life of which he is an integral part. There can be no question that the sense of this duty which mastery of the earth gives or should afford is to be one of the moral gifts of modern learning. So long as men considered themselves to be accidents on the earth, imposed upon it by the will of a Supreme Being, but in nowise related in origin and history to the creatures amid which they dwelt, it was natural that they should exercise a careless and despotic power over their subjects. Now that it has been made perfectly clear that we have come forth from the maze of the lower life, that all these tenants of the wilderness are sharers in the order which has brought us to our estate, and that each one of them, plant and animal alike, is the record of the impulses which lead beings upward, we can no longer keep the old careless attitude. We are compelled to deal with the organic hosts as we deal with the creatures of our folds and fields. We have to look upon them all as a member of the great household of man, made such by the intellectual conquest of the world to which he has attained. We may trust the sense of this large duty to extend abroad under the influences which have developed it in the minds of a few men, or we may hasten its development by a propaganda such as is carried on by the societies for the prevention of cruelty to [264] animals. If this latter course is taken the teaching should be on a higher plane than that which we have yet had from those generally admirable associations. Bad as is the ill treatment of domesticated animals, the suppression of that evil will not bring us materially nearer the true attitude that we need to assume in face of our responsibilities to the natural world. We need to see the greatness of the responsibility which has been imposed upon us by the action of the guiding power that has made us lords of the earth.

[265]

www.ingramcontent.com/pod-product-compliance
Lightning Source LLC
Chambersburg PA
CBHW031619210526
45464CB00004B/1662